DIRECTORY OF AQUACULTURE SCIENTIST

Compiled by
Akhilesh Kumar Singh

2000
DAYA PUBLISHING HOUSE
DELHI-110035

AKHILESH KUMAR SINGH (b. 1963)

ISBN 81-7035-219-3

Published by : **Daya Publishing House**
1123/74, Deva Ram Park
Tri Nagar, Delhi-110 035
Phone : 7103999, 7199029
Fax : (011) 7198902
E-mail : info@dayabooks.com
Website : www.dayabooks.com

Laser Composed by : **Vaishnav Graphics & Systems**
New Delhi - 110 055
Phone : 3533756

Printed at : **Chawla Offset Printers**
Delhi-110052

PRINTED IN INDIA

Dedicated to My Sons

Alok & Anup

BIRSA AGRICULTURAL UNIVERSITY
RANCHI - 834 006

Dr. R.P. Roy Sharma
M.Sc.(Agron.), Ph.D.(IARI), FISA, FISPR
VICE-CHANCELLOR

Date : 28.11.99

FOREWORD

Aquaculture is a multidisciplinary subject which is growing very fast. Many scientists are working all over the world to have more production through aquaculture. The present work "Directory of Aquaculture Scientist" is a useful effort by Dr. A.K. Singh of this University. It will be useful for the Scientists, Researchers and Non-Governmental organisation working in aquaculture and needs consultancy services.

I wish all success and congratulate Dr. Singh for such a nice effort.

Ranchi
Date : 28/11/99

R.P. Roy Sharma

P.O.-KANKE, RANCHI - 834 006, BIHAR, INDIA, TEL.: 0651-455500 (O); 455866, 455623 (R)
CABLE : BIRSAAGUNIV, VSAT : 32, FAX : 0651-455850, E-mail : bau@bitsmart.com

PREFACE

This is an attempt to have first Directory of Aquaculture Scientists. The aim of this Directory is to make the scientist closer and provide a ready references to NGO and other organisations who needs consultancy services in Aquaculture.

The content of the Directory has been collected from the response of a questionnaire which was sent to most of the Aquaculture Institutes and individuals. Some information has been collected from the Directory available of different institutions and associations. This Directory has two sections, Section-1 contains all the entries in alphabetical order (last name) and a reference number has been provided. In the section-2 references has been given on the basis of major and minor field of interest of scientists.

The things are changing every day, so there is a chance for correction of addresses etc. If there is any change/correction please bring it to the notice, so that it can be corrected in the next issue.

This Directory is a small collection only, still more names have to be included, if you feel interested please fill up the form available in the last page of the book and send to us.

I am very much thankful to Anil Mittal of Daya Publishing House, Delhi for his keen interest in the Directory. I am also thankful to all who helped me specially my family members who kept me free from household duties and encouraged me for this work.

I hope this piece of work will be useful to researchers and organisations who need consultancy services in Aquaculture.

There is scope for improvement, your suggestions for improvement will be highly appreciated.

18th August, 1999 **Akhilesh Kumar Singh**

1. Mr. Sotero M. **Aban**
Instructor III, College of Fisheries
Pangasinan State University, Binmaley
Pangasinan 2417, Philippines
Subject of Interest
Major : Fresh water Aquaculture, Nutrition, water quality
Minor : Genetics, Integrated fish farming, tilapias

2. Mr. Tereso A. **Abella**
Research Biologist, Freshwater
Aquaculture Center, Central Luzon
State University Munoz, Nueva Ecija
3120, Philippines
Contact No.(s) : (Fax) 3017323
Subject of Interest
Major : Biotechnology genetics, quantitative genetics.

3. Ms. Belen O. **Acosta**
Research Associate, ICLARM
MCPO Box 2631, 0718 Makati
Metro Manila, Philippines
Contact No.(s) : (Tel.) 8189283, 8180466, 8175255
(Fax) 8163183
Subject of Interest
Major : Freshwater aquaculture
Minor : Bivalves, tilapias

4. Dr. Oluwakayode A. **Adekeye**
Deputy Director/Head of Fisheries
Ministory of Agriculture and Natural Resources,
PMB 1383, Ilorin, Kwara State, Nigeria
Contact No.(s) : (Tel.) 220063
Subject of Interest
Major : Development, extension, policy
Minor : Conservation, environmental studies, information.

5. Mr. Joseph A. **Adetayo**
Senior Research Officer,
Nigerian Institute for Oceanography and Marine Research,
PMB 12790, Victoria Island Lagos, Nigeria
Contact No.(s) : (Tel.) 617530
Subject of Interest
Major : Brackishwater aquaculture, crustaceans, induced spawning.
Minor : Computer methods, hachery technology software

6. Dr. Innocent **A. Adikwu**
Lecturer, Department of Biological Science,
Bayero University, PMB 3011,
Kano, Nigeria
Subject of Interest
Major : Feed technology
Minor : Nutrition

7. Dr. J.F. **Agnese**
Centre de recherches oceanologiques
BPV 18, Abidjan, Cote d'Ivoire
Contact No.(s) : (Tel.) 355014
Subject of Interest
Major : Genetic resources, population genetics, taxonomy.
Minor : Catfishes, stock assessment, tilapias

8. Dr. Narayan K. **Agrawal**
Aquaculture Engineer
Orissa Shrimp Seed Production, Supply
and Research Center, Gopalpur, On-Sea
Dist, Ganjam, Orissa, India
Contact No.(s) : (Tel.) 8165, 8160
Subject of Interest
Major : Brackishwater aquaculture, engineering, hatchery technology.
Minor : Farming management

9. Ms. Liza Q. **Agustin**
Research Assistant, ICLARM
MCPO Box 2631, 0718 Makati
Metro Manila, Philippines

Contact No.(s) : (Tel.) 8189283, 8180466, 8175255
(Fax) 8163183

Subject of Interest

Major : Physiology, population genetics

Minor : Freshwater aquaculture

10. Dr. Muohtar **Ahmad**
Chief of Aquaculture Laboratory/Lecturer
Fakultas Perikanan, University Riau
J1. Pattimura No. 9, Pekanbaru, Riau
28131, Indonesia

Subject of Interest

Major : Behavior, economics, integrated farming systems

Minor : Hatchery technology, marine aquaculture, physiology

11. Mr. Khan K.U. **Ahmed**
Scientific Officer, Riverine Sub-station
Fisheries Research Institute
PO Box 8, Rangamati 4500, Bangladesh
Contact No.(s) : (Tel.) 2159

Subject of Interest

Major : Limnology, population dynamics, stock assessment.

Minor : Cages, phytoplankton, water quality.

12. Dr. Mahfuzuddin **Ahmed**
Technical Coordinator, ICLARM Bangladesh
Office, House 20, Rd. 9/A,
Dhanmondi R/A
Dhaka 1209, Bangladesh
Contact No.(s) : (Tel.) 812466, 817345
(Fax) 884968, 883416

Subject of Interest

Major : Economics, integrated farming systems, resources economics

Minor : Farming management, farming systems research, socioeconomics.

13. Mr. Yemi **Akegbejo-Samsons**
Consultant, PO Box 637, Akure, Ondo
State, Nigeria

Subject of Interest
Major : Freshwater aquaculture
Minor : Brackishwater aquaculture, integrated farming systems

14. Ms. Jubaida N. **Akhter**
Scientific Officer, Fisheries Research
Institute, Riverine Station, Baburhat,
Chandpur 3602, Bangladesh
Contact No.(s) : (Tel.) 3482, 3407
Subject of Interest
Major : Cages, nutrition, pens
Minor : Computer methods, feed technology, freshwater aquaculture

15. Mr. Syed Arif M. **Al-Arabi**
Principal Investigator, Fisheries
Research Institute, Riverine Station
Baburhat, Chandpur 3602, Bangladesh
Contact No.(s) : (Tel.) 3016, 3407
Subject of Interest
Major : Environmental impact, pollution, water quality
Minor : Genetics, integrated farming systems, pesticides

16. Dr. Ahmed H. **Al-Harbi**
Institute of Natural Resources
and Environmental Research
King Abdulaziz City for Science and
Technology, PO Box 6086, Riyadh 11442,
Saudi Arabia
Contact No.(s) : (Tel.) 401590
Subject of Interest
Major : Diseases, immunology, pathology
Minor : Microbiology, parasities

17. Mr Oluwasegun W. **Alegbeleye**
Senior Research Officer
Nigerian Institute for Oceanography
and Marine Research,
PMB 12729, Victoria Island, Lagos,
Nigeria

Subject of Interest

Major : Freshwater aquaculture, induced spawning integrated farming systems

Minor : Catfishes, marine aquaculture, tilapias

18. Dr. Ahyaudin **B. Ali**
Associate Professor, School of Biological Sciences,
University Sains Malaysia
11800 Minden, Penang, Malaysia
Contact No.(s) : (Tel.) 877888
(Fax) 871526

Subject of Interest

Major : Farming systems research, freshwater aquaculture, rice-fish farming systems

Minor : Limnology, pond dynamics, reservoirs.

19. Dr. Sayed M. **Ali**
Red Sea Fisheries Research Centre,
PO Box 730, Port Sudan, Sudan

Subject of Interest

Major : Bivalves, coral reefs, marine aquaculture

Minor : Environmental studies, giant clams, statistics

20. Ms. Peninah A. **Aloo**
Lecturer, Zoology Department,
Kenyatta University, PO Box 43844,
Nairobi, Kenya
Contact No.(s) : 810901

Subject of Interest

Major : Diseases, parasites, pathology

Minor : Freshwater aquaculture, limnology, tilapias

21. Golam Mahbub **Alam**
Fisheries Research Institute,
Riverine Station, Chandpur,
3602, Bangladesh

22. Mr. Aggrey J. **Ambali**
Lecturer, Biology Department, University
Malawi Chancellor College, PO Box 280,
Zomba, Malawi

Contact No.(s) : (Tel.) 522222
(Fax) 522046
Subject of Interest
Major : Broodstock, population genetics, quantitative genetics
Minor : Seed production, tilapias

23. Mr. Francis K.Y. **Amevenku**
Research Officer, Institute of Aquatic
Biology, PO Box 38, Achimota, Ghana
Contact No.(s) : (Tel.) 775511, 777655
Subject of Interest
Major : Genetics, pond management, reproductive physiology

24. Mr. Stephen **Amisah**
Lecturer, Institute of Renewable Natural Resources
Department of Freshwater Fisheries and
Watershed Management, University of
Science and Technology, Kumasi, Ghana
Contact No.(s) : (Tel.) 5354
Subject of Interest
Major : Freshwater aquaculture, genetics, induced spawning
Minor : Ecology, limnology, marine aquaculture

25. Dr. El Abed **Amor**
Maitre de conferences
Ecole superieure d'horticulture et
d'elevage 4042 Chott Meriam,
Sousse, Tunisia
Contact No.(s) : (Tel.) 48134
(Fax) 48275
Subject of Interest
Major : Environmental impact
Minor : Pollution, reproductive physiology

26. Mr. Ramanujam **Ananthanarayanan**
Deputy Director of Fisheries, Salt
Road, Nagappattinam, Tamil Nadu, India
Contact No.(s) : (Tel.) 2321

Subject of Interest
Major : Freshwater aquaculture
Minor : Brackishwater aquaculture, extension

27. Ms. Merlina N. **Andalecio**
University Research Associate II
Institute of Fisheries Policy and Development
Studies, College of Fisheries
University of the Phillippines in the Visayas,
Miag-ao, Iloilo 5023, Philippines
Subject of Interest
Major : Bioenergetics, environmental studies, taxonomy
Minor : Computer methods, ecophysiology, modeling

28. Mr. Alexander J. **Anderson**
Lecturer in Biochemistry, School of Life
Science Queensland University of Technology,
GPO Box 2434, Brisbane 4001, Australia
Contact No.(s) : (Tel.) 8642984
(Fax) 8641534
Subject of Interest
Major : Biochemistry, freshwater aquaculture, nutrition
Minor : Brackishwater aquaculture, feed technology, finfish

29. Dr. **Ang Kok-Jee**
Professor of Aquaculture
Faculty of Fisheries and Marine Science
University Pertanian Malaysia, 43400
UPM Serdang Selangor Darul Ebsan,
Malaysia
Contact No.(s) : (Tel.) 9486101
(Fax) 9482507
Subject of Interest
Major : Farming systems research, freshwater aquaculture
Minor : Wastewater (sewage) reuse

30. Mr. Kambamba N. **Anson**
Directeur, Centre rural pour le development
integre de Bonga-Yasa BP 16.383
Kinhasa 1, Zaire

Subject of Interest
Major : Catfishes, development, freshwater aquaculture
Minor : Integrated farming systems, pond management, tilapias

31. Mr. Evi M. **Antoniou**
28 Asklipiou St., Katholikis, Limassol,
Cyprus
Contact No.(s) : (Tel.) 27608
(Fax) 37069
Subject of Interest
Major : Marine aquaculture, nutrition, pathology
Minor : Crustaceans, environmental impact, tilapias

32. Mr. J. Davies **Anuta**
Deputy Provost
Federal College of Fisheries and Marine Technology PMB 80063, Victoria Island,
Lagos, Nigeria
Contact No.(s) : (Tel.) 613820
Subject of Interest
Major : Nutrition, tilapias, water quality
Minor : Production systems

33. Ms. Patricia E. **Anyanwu**
Research Officer
Nigerian Institute for Oceanography and Marine Research
PMB 12729, Victoria Island, Lagos, Nigeria
Contact No.(s) : (Tel.) 617530
Subject of Interest
Major : Genetics, seed production, tilapias
Minor : Brackishwater aquaculture, farming systems research, genetic resources

34. Mr. K.K. **Appukuttan**
Scientist, Regional Center of Central Marine Fisheries Research Institute
West Hill, PO Calicut - 673005, India
Contact No.(s) : (Tel.) 56651

Subject of Interest
Major : Bivalves, hatchery technology, marine aquaculture
Minor : Brackishwater aquaculture, seed production, stock assessment

35. Mr. Zainal **Arifin**
Researcher, BALITBANG SDL-LIPI
Guru-guru, Poka, Ambon 97233, Indonesia
Contact No.(s) : (Tel.) 69352
Subject of Interest
Major : Conservation, farming management, reproductive physiology
Minor : Coral reefs, quantitative genetics

36. Dr. Venkatesan **Arul**
Centre for Biotechnology
Department of Life Sciences
Pondichery University, Kalapet
Pondichery 605014, India
Subject of Interest
Major : Bioenergetics, biotechnology, endocrinology
Minor : Induced spawning, nutrition, reproductive physiology

37. Mr. Ambrose **Asamoah**
Assistant Research Officer Institute
of Aquatic Biology PO Box 38,
Achimota, Ghana
Contact No.(s) : (Tel.) 775511
(Fax) 777655
Subject of Interest
Major : Freshwater aquaculture, induced spawning, rich-fish farming systems
Minor : Brackishwater aquaculture, hatchery technology, integrated farming systems

38. Ms. Grace L. **Asheri**
Research Fisheries Officer, Regional
Fisheries Office PO Box 3012,
Arusha, Tanzania

Subject of Interest
Major : Catfishes, freshwater aquaculture, tilapias
Minor : Diseases, nutrition

39. Ms. Jema N. **Awa**
Research Officer, Nigerian Institute
for Oceanography and Marine Research
PMB 12729, Victoria Island, Lagos
Nigeria
Contact No.(s) : (Tel.) 617530
Subject of Interest
Major : Growth, pathology, stock assessment
Minor : Computer methods, diseases, parasities

40. Dr. Olajide A. **Ayinla**
Centre Manager, African Regional
Aquaculture Centre PMB 5122, Port
Harcourt, Rivers State, Nigeria
Subject of Interest
Major : Farming systems research, feed technology, hatchery technology
Minor : Brackishwater aquaculture, extension training

41. Dr. Subbanna **Ayyappan**
Director, Central Institute of Freshwater
Aquaculture, Kausalyaganga,
Bhubaneswar-751002, India
Contact No.(s) : (Tel.) 0674-463421, 463446, (R) 430796
(Fax) 0674-463407
Subject of Interest
Major : Fisheries
Minor : Aquaculture
Subject of Consultancy
1. Aquatic microbiology and Ecology.
2. Waste water treatment through Aquaculture.
3. Carp culture systems.

42. Dr. Richard Donald **Braley**
Aquasearch
Neny Bay Magnetic Island,
Queensland 4819, Australia

Contact No.(s) : (Tel.) 61747785096
(Fax) 61747785096

Subject of Interest

Major : Tropical Aquaculture, Shell fish (Giant clams, Pearls oysters, trochas) and Aquarium-cultured species.

Minor : Marine biology/ecology, Fisheries, stock assessment, marine science, training.

Subject of Consultancy

Tropical marine aquaculture.
Hatchery design and set up
Training hatchery and farm techniques
Trouble shooting in hatchery and farm.
Sea water system design and onsite assistance.
Site survey for hatchery and field grow out.
Environmental and field stock surveys.
Recirculation and biofiltration.
Marine aquarium fish and invertebrates culture

43. Mr. John D. **Balarin**
Project Advisor,
DANIDA Technical Assistance
Oxbow Lake Small Scale
Fishermen Project,
Box 12, Jessore 7600,
Bangladesh

Contact No.(s) : (Tel.) 4156

Subject of Interest

Major : Freshwater aquaculture, integrated farming systems, tilapias

Minor : Broodstock, carps, farming systems research

44. Ms. Arcangel A. **Balicanta**
Senior Aquaculturist
Department of Agriculture
Laoag City Ilocos Norte 2900,
Philippines

Subject of Interest

Major : Seaweeds

Minor : Bivalves, freshwater aquaculture

45. Mr. Joseph J. **Baloyi**
Research Officer, Henderson Research
Station, PB 2004, Mazowe, Zimbabwe
Subject of Interest
Major : Nutrition

46. Mr. Edwin F. **Bartholomew**
Aquaculture Consultant, 620 Laie
Drive, Makawao, HI 96768, USA
Contact No.(s) : (Tel.) 5729979
Subject of Interest
Major : Freshwater aquaculture
Minor : Water quality, zooplankton

47. Dr. Devin **Bartley**
Fishery Resources Officer, FIRI-FAO
Via delle Terme di Caracalla, 00100
Rome, Italy
Contact No.(s) : (Tel.) 57976470
(Fax) 5120330
Subject of Interest
Major : Genetic resource
Minor : Conservation, population genetics

48. Mr. N. **Basavaraja**
Assistant Professor, University of
Agricultural Sciences, College of Fisheries,
Managalore - 575002, Karnataka (India)
Subject of Interest
Major : Freshwater aquaculture
Minor : Biotechnoloy, nutrition

49. Mr. Matas N. **Batao**
Dean / Associate Professor IV
College of Fisheries, Mindanao State
University, Maguindanao 9601,
Philippines
Subject of Interest
Major : Brackishwater aquaculture, freshwater aquaculture
Minor : Rice-fish farming systems

50. Dr. Antonio M. **Bautista**, Sr.
Associate Professor, College of Fisheries,
Pangasinan States University Binmaley,
Pangasinan 2417, Philippines
Subject of Interest
Major : Freshwater aquaculture
Minor : Integrated farming systems, tilapias

51. Dr. Leslie L. **Behrends**
Regional Manager, Agriculture Institute,
PO Box 1010, Muscle Shoals,
AL 35660, USA
Contact No.(s) : (Tel.) 3863488
(Fax) 3862284
Subject of Interest
Major : Genetics, hatchery technology, tilapias
Minor : Farming systems research, integrated farming systems, pond dynamics

52. Ms. Carmelita A. **Belda**
Research Associate,
Zoology Department,
James Cook University Townsville,
Queensland 4811, Australia
Contact No.(s) : (Tel.) 814853, 814563
Subject of Interest
Major : Bivalves, giant clams, marine aquaculture
Minor : Nutrition, physiology, water quality

53. Dr. Hans Bl **Bentsen**
Senior Scientist,
Institute of Aquaculture Research,
Agricultural University of Norway,
PO Box 10, N-1432, AS,
Norway
Contact No.(s) : (Tel.) 949507
(Fax) 949502
Subject of Interest
Major : Quantitative genetics
Minor : Finfish

54. Dr. John A.H. **Benzie**
Senior Research Scientist
Australian Institute of Marine Science
PMB No. 3 Townsville, Queensland 4810,
Australia
Contact No.(s) : (Tel.) 789211
(Fax) 725852
Subject of Interest
Major : Crustaceans, genetics, marine aquaculture
Minor : Biotechnology, conservation, environmental impact

55. Dr. Garry **M. Bernacsek**
Fisheries Specialist, Northeast
Regional Project, CIDA, PO Box 6096,
Gulshan, Dhaka, Bangladesh
Contact No.(s) : (Tel.) 882823
(Fax) 883097
Subject of Interest
Major : Environmental impact, policy, stock assessment
Minor : Conservation, habitat restoration, resource economics

56. Mr. Michel **Berube**
Charge de projet, Hydro-Quebec
500 Sherbrooke O. Montreal, Quebec
Canada H3A 3C6
Contact No.(s) : (Tel.) 2862094
(Fax) 2867614
Subject of Interest
Major : Environmental studies
Minor : Limnology, pond management

57. Dr. Malcolm C.M. **Beveridge**
Senior Lecturer, Institute of Aquaculture
University of Stirling,
Stirling FK9 4LA, UK
Contact No.(s) : (Tel.) 467885
(Fax) 472133
Subject of Interest
Major : Ecology, environmental impact, foodchains
Minor : Cages, detritus, resource economics

58. Mr. Rai M. **Bhowmick**
Managing Director,
Fisheries Nautrewealth Development Corporation Ltd.
Camp Office Ram Nagar,
PO Bashra Murshidabad,
West Bengal, India
Subject of Interest
Major : Freshwater aquaculture, hatchery technology, induced spawning.

59. Ms. Mary Ann P. **Bimbao**
Research Associate,
ICLARM
MCPO Box 2631, 0718 Makati
Metro Manila, Philippines
Contact No.(s) : (Tel.) 8189283, 8175255, 8180466
(Fax) 8163183
Subject of Interest
Major : Economics, integrated farming systems, rice-fish farming systems
Minor : Farming systems research, resources economics, socio-economics

60. Dr. Bhaskar R. **Bollapragada**
Technical Officer,
Rallis India Ltd.
21 D.S. Marg, Bombay 400001, India
Contact No.(s) : (Tel.) 846311
Subject of Interest
Major : Brackishwater aquaculture, farming management, pond management
Minor : Economics, extension, post-harvest technology

61. Ms. Grace B. **Brizuela**
Assistant Professor
3, College of Fisheries, Bicol University,
Tabaco, Albay 4511,
Philippines
Contact No.(s) : (Tel.) 627

Subject of Interest

Major : Brackishwater aquaculture, limnology, water quality

Minor : Environmental studies, extension, freshwater aquaculture, integrated farming systems, milkfish, pond management, rice-fish farming systems, site selection, statistics, tilapias, water systems

62. Mr. Alan C. **Brooks**
Research Officer,
PO Box 700,
Mzuzu, Malawi
Contact No.(s) : (Tel.) 332794
(Fax) 335051

Subject of Interest

Major : Freshwater aquaculture, reservoirs, tilapias

Minor : Extension, integrated farming systems, stock assessment

63. Dr. Randall E. **Brummett**
Director,
ICLARM/GTZ Africa Aquaculture
Project, PO Box 229, Zomba, Malawi
Contact No.(s) : (Tel.) 531274, 531215
(Fax) 528103

Subject of Interest

Major : Integrated farming systems, quantitative genetics, Tilapias

Minor : Freshwater aquaculture, population genetics, reproductive physiology

64. Ms. Ma. Gladys G. **Buenavista**
Ph.D. Candidate,
Department of Sociology,
McBryde Hall Virginia
Polytechnic Institute and State University,
Blacksburg, VA 24061-0137, USA
Contact No.(s) : (Tel.) 2316878
(Fax) 2313860

Subject of Interest

Major : Environmental studies

65. Mr. **Bui Thi Lang**
Head, Department of Mangrove
79 Trouong Dinh, QI, Ho Chi Minh City,
Vietnam
Contact No.(s) : (Tel.) 297001
(Fax) 324400
Subject of Interest
Major : Brackishwater aquaculture, ecology, habitat restoration
Minor : Crustaceans, environmental impact, farming management

66. Ms. Louise G. **Buttle**
Ph.D. Student
Humberside International Fisheries
Institute, University of Hull,
HUG 7RX, UK
Subject of Interest
Major : Catfishes, physiology, water quality
Minor : Ecophysiology, freshwater aquaculture, integrated farming systems, rural sociology

67. Mr. Joyson K. **Buya**
Ph.D. Student, School of Biological
Sciences, University College of Swansea
Singleton Park, Swansea SAI 8PP, UK
Subject of Interest
Major : Genetics
Minor : Biodiversity, freshwater aquaculture

68. Dr. Philip O.J. **Bwathondi**
Director General,
Tanzania Fisheries Research Institute,
PO Box 9750
Dar Es Salaam, Tanzania
Contact No.(s) : (Tel.) 73563
Subject of Interest
Major : Broodstock, cages, marine aquaculture
Minor : Brackishwater aquaculture

69. Sumen **Chakrabarti**
C/o Dilip Kr. Chakrabarti
East Shib Nagar
P.O Agaratala College
Tripura (W)-799004 India
Subject of Interest
Major : Fishery and Biology of fresh water prawns. Freshwater Prawn culture.
Minor : Ecology and fisheries of wet lands of India and abroad.
Subject of Consultancy
Identification characters, morphology and life cycles of new species of the freshwater shrimp of the genus caridian. Fishery and biology of Macrobrachium spp. of the wet land ecosystems of India and abroad. Ecology of wet land ecosystems of India and abroad.

70. Ms. Arsenia G. **Cagauan**
Assistant Professor,
Freshwater Aquaculture Center,
Central Luzon State University,
Munoz, Nueva Ecija 3120,
Philippines
Contact No.(s) : (Tel.) 3017323
Subject of Interest
Major : Freshwater aquaculture
Minor : Water Quality

71. Dr. Jorge O. **Calderon**
PO Box 09-01-4519,
Guayaquil, Ecuador
Contact No.(s) : (Tel.) 765153
Subject of Interest
Major : Brackishwater aquaculture, modeling, pond dynamics
Minor : Growth, induced spawning, marine aquaculture

72. Ms. Ma. Luisa **Carranza Arauna**
Raimundo Fernandez-Villaverde,
No. 40-342, Madrid, Spain

Subject of Interest
Major : Marine aquaculture, pathology
Minor : Physiology

73. Mr. Arturo **Chacon-Torres**
Head of Aquatic Biology Laboratory
Edificio "R" Planta Baja
Escuela de Biologia, Ciudad Universitaria,
Morelia, Michoacan, Mexico
Contact No.(s) : (Tel.) 167412
(Fax) 144634
Subject of Interest
Major : Ecology, limnology, water quality
Minor : Computer methods, freshwater aquaculture, site selection

74. Mr. Momodou A. **Cham**
Assistant Director, Fisheries Department,
6 Marina Parade, Banjul, Gambia
Contact No.(s) : (Tel.) 28727, 27339
(Fax) 29701
Subject of Interest
Major : Bivalves, brackishwater aquaculture, crustaceans
Minor : Hatchery technology, marketing, socioeconomics

75. Dr. Gunindra N. **Chattopadhyay**
Soil Scientist, Institute of Agriculture
Visva Bharati University, Sriniketan
731236, (India)
Subject of Interest
Major : Pond management, rice-fish farming systems, waste-fed
Minor : Environmental studies, farming management, farming systems research

76. Dr. Rikhi S. **Chauhan**
Assistant Profressor,
Department of Aquaculture, College of Fisheries,
G.B. Pant University of Agriculture and
Technology, Pantnagar 263149, U.P., India

Subject of Interest
Major : Biotechnology, freshwater aquaculture, pathology
Minor : Farming management, induced spawning, site selection

77. Mr. **Cheah Sin-Hock**
C/o Dr. Chan L. Lee
School of Biological Sciences
Northern Territory University
PO Box 40167, Casuarina,
Northern Territories 0811, Australia
Contact No.(s) : (Tel.) 466666
(Fax) 466690, 270612
Subject of Interest
Major : Induced spawning
Minor : Carps, catfishes, hatchery technology

78. Mr. A. **Chellam**
Scientist, Tuticorin Research Centre of
Central Marine Fisheries Research Institute,
90 North Beach Rd.,
Tuticorin 628001, India
Contact No.(s) : (Tel.) 21515
Subject of Interest
Major : Bivalves, hatchery technology, marine aquaculture, pearl culture

79. Dr. **Chen Liqiao**
Vice-Professor, Department of Biology,
East China Normal University
Shanghai 200062, China
Contact No.(s) : (Tel.) 2577577
Subject of Interest
Major : Crustaceans, freshwater aquaculture, nutrition
Minor : Biotechnology, finfish, genetics

80. Mr. **Chen Zhen-Ran**
Professor, Laboratory of
Marine Biology, South China Sea
Institute of Oceanology Academia
Sinica, 164 West Xin Gang Rd.,
Guangzhou 510301, China
Contact No.(s) : (Tel.) 4451335 to 510
(Fax) 4451672
Subject of Interest
Major : Marine aquaculture, stock assessment
Minor : Brackishwater aquaculture, ecology

81. Mr. Redson J.K. **Chikafumbwa**
Research Associate III
ICLARM/GTZ Africa Aquaculture Project
PO Box 229, Zomba, Malawi
Contact No.(s) : (Tel.) 531274, 531215
(Fax) 528103
Subject of Interest
Major : Freshwater aquaculture, integrated farming systems, tilapias
Minor : Catfishes, nutrition, pond management, rich-fish farming systems

82. Dr. Supranee **Chinabut**
Fish Pathologist
Aquatic Animal Health Research Institute,
Kasetsart University Campus
Bangkhen 10900, Bangkok, Thailand
Contact No.(s) : (Tel.) 5794122
Subject of Interest
Major : Diseases, parasites, pathology
Minor : Immunology

83. Dr. Satish C. **Choy**
Senior Lecturer, Department of Biology
Universiti Brunei Darussalam
Gandong 3186, Brunei Darussalam
Contact No.(s) : (Tel.) 427001
(Fax) 427003

Subject of Interest
Major : Ecophysiology
Minor : Crustaceans, finfish

84. Mr. Mikkel S. **Christensen**
Consultant, Frankfurter Str. 184607
D-6078 Neu-Isenburg, Germany
Contact No.(s) : (Tel.) 34541
Subject of Interest
Major : Environmental impact, freshwater aquaculture, integrated farming systems
Minor : Brackishwater aquaculture, conservation, marine aquaculture

85. Mr. Antonio V. **Circa**
Assistant Professor, Freshwater aquaculture
Center, Central Luzon State University,
Munoz, Nueva Ecija 3120,
Philippines
Contact No.(s) : (Tel.) 3017323
Subject of Interest
Major : Freshwater aquaculture, integrated farming systems, waste-fed
Minor : Pond dynamics, rice-fish farming systems, water quality

86. Mr Adou **Cisse**
Chef du department aquaculture
Centre de recherches oceanologiques
BP V 18 Abidjan, Cote d'Ivoire
Contact No.(s) : (Tel.) 355014, 355880
(Fax) 351155
Subject of Interest
Major : Growth, nutrition, tilapias
Minor : Farming systems research

87. Dr. David **Coates**
FAO Field Office
Box 305, Madang, Papua New Guinea
Contact No.(s) : (Tel.) 823011
(Fax) 823306

Subject of Interest
Major : Fisheries enhancement
Minor : Environmental impact, socioeconomics, stock assessment

88. Dr. Andre G. **Coche**
Via Adolfo Vigorelli, 7, Rome 00144,
Italy
Contact No.(s) : (Tel.) 5295977
Subject of Interest
Major : Development, extension, freshwater aquaculture
Minor : Cages, rice-fish farming systems, tilapias

89. Mr. Robert D. **Cordover**
Department of Conservation, Forests
and Lands PO Box 41, East Melbourne,
Victoria 3302, Australia
Contact No.(s) : (Tel.) 8034691
(Fax) 4124623
Subject of Interest
Major : Brackishwater aquaculture
Minor : Hatchery technology, pond management

90. Dr. Rhodora A. **Corrales**
Associate Professor, Marine Science
Institute University of the Philippines,
Diliman 1101, Quezon City,
Philippines
Contact No.(s) : (Tel.) 986976
Subject of Interest
Major : Growth, physiology, seaweeds

91. Dr. Barry A. **Costa-Pierce**
Director of Research and Development
Mauna Loa Research Institute, Inc., 6
Hale Ohia Rd. PO Box 1050, Volcano,
HI 96785, USA
Contact No.(s) : (Tel.) 9678607
(Fax) 9858608

Subject of Interest
Major : Freshwater aquaculture, integrated farming systems, reservoirs
Minor : Foodchaines, pond dynamics, production systems

92. Mr. Marcelo **Costero**
Director, Laboratorio de Ciencias
Marinas, Apartado Postal No. 3,
Barra de Navidad, Jalisco 48987, Mexico
Contact No.(s) : (Tel.) 33370130
(Fax) 33370168

Subject of Interest
Major : Crustaceans, hatchery technology, marine aquaculture
Minor : Freshwater aquaculture, tilapias, water quality

93. Dr Peter **Coutteau**
Research Assistant, Faculty of Agricultural
Sciences, Laboratory of Aquaculture
and Artemia Centre
Rozier 44, B-9000 Ghent, Belgium
Contact No.(s) : (Tel.) 643754
(Fax) 644193

Subject of Interest
Major : Bivalves, nutrition, seed production
Minor : Feed technology, hatchery technology, zooplankton

94. Dr. Ian G. **Cowx**
Senior Lecturer
Humberside International Fisheries
Institute, University of Hull, Hull
HU6 7RX, UK
Contact No.(s) : (Fax) 470129

Subject of Interest
Major : Extension, freshwater aquaculture, integrated farming systems
Minor : Computer methods, environmental impact, farming management.

95. Dr. Donatella **Crosetti**
Researcher, Instituto Centrale per la
Ricerca Scientificae Technologica
Applicata alla Pesca Marittima
P.zza ISTRIA 3-00198, Rome, Italy
Contact No.(s) : (Tel.) 8607558
(Fax) 8088326

Subject of Interest
Major : Genetic resources, genetics, population genetics
Minor : Integrated farming systems, lagoons, tilapias

96. Dr. Emmanuel M. **Cruz**
Research Scientist
Mariculture and Fisheries Department
Kuwait Institute for Scientific Research,
PO Box 1638, 22017 Salmiyah
Kuwait
Contact No.(s) : (Tel.) 4816237
(Fax) 4846891

Subject of Interest
Major : Integrated farming systems, nutrition, tilapias
Minor : Cages, marine aquaculture

97. Mr. Flinn **Curren**
Aquaculture Consultant, PO Box 1517
Kolonia Pohnpei, FM 96941, Micronesia
Contact No.(s) : (Tel.) 5886, 2795
(Fax) 5997

Subject of Interest
Major : Giant clams, marine aquaculture, stock assessment
Minor : Environmental studies, socioeconomics

98. Prof. Nelson Magno **Doren**
Faculty of Marine Sciences
King Abdulaziz University
P.O. Box 1540
Obhur Jeddah 21441
Kingdom of Saudi Arabia
Contact No.(s) : (Tel.) (2) 699-2203
(77) 793-6951 (R)
(Fax) 699-2092

Subject of Interest

Major : Fish pond layout and design. Finfish culture crustacean culture

Minor : Marine Aquarium installation and management, Sea weed culture.

Subject of Consultancy

Fish pond lay outing and design and management, Shrimp and prawn hatchery operation, and management. Commercial production of tilapia in tanks and other artificial culture systems. Construction, installation and management of marine aquarium.

99. Mr. Stephen **Dadzie**

Chairman, Department of Zoology
University of Nairobi
PO Box 30197, Nairobi, Kenya
Contact No.(s) : (Tel.) 442316
(Fax) 336885

Subject of Interest

Major : Freshwater aquaculture, reproductive physiology, training

Minor : Ecology, reservoirs, seed production

100. Mr. Jens Peter Tang **Dalsgaard**

Assistant Scientist, ICLARM
MC PO Box 2631, 0718 Makati
Metro Manila, Philippines
Contact No.(s) : (Tel.) 8189283, 8180466
(Fax) 8163183

Subject of Interest

Major : Ecology, farming systems research, modeling

Minor : Environmental impact, integrated farming systems, rice-fish farming systems

101. Mr. Paul J. **Dalzell**

Inshore Fisheries Scientist, South
Pacific Commission BP D5 Noumea,
New Caledonia
Contact No.(s) : (Tel.) 262000
(Fax) 263818

Subject of Interest
Major : Coral reefs, growth, stock assessment
Minor : Brackishwater aquaculture, modeling, policy

102. Mr. Alfred L. **Dassah**
Lecturer/Head of Department
Institute of Renewable Natural Resources
University of Science and Technology,
Kumasi, Ghana
Subject of Interest
Major : Freshwater aquaculture, integrated farming systems
Minor : Water systems.

103. Dr. F. Brian **Davy**
Associate Director, IDRC
PO Box 8500, Ottawa, Ontario, Canada
KIG 3H9
Contact No.(s) : (Tel.) 2366163
L : 2209
(Fax) 5677749
Subject of Interest
Major : Curriculum development, farming systems research, marine aquaculture

104. Dr. Sena S. **De silva**
Coordinator, Post-Graduate Courses
Faculty of Aquatic Science, Deakin
University PO Box 423, Warrnambool,
Victoria 3280, Australia
Contact No.(s) : (Tel.) 633527
(Fax) 633531
Subject of Interest
Major : Curriculum development, nutrition, reserviors
Minor : Feed technology, freshwater aquaculture, marine aquaculture

105. Mr Padmakar V. **Dehadrai**
Deputy Director General
Indian Council of Agricultural Research
Ministry of Agriculture, Rm. 103, Krishi Bhawan,
Dr. Rajendra Prasad Rd., New Delhi-110001, India

Contact No.(s) : (Tel.) 672385, 382713
(Fax) 387293
Subject of Interest
Major : Aquaculture development
Minor : Ecology, farming management

106. Dr. Catalino R. **Dela Cruz**
Consultant, ICLARM
MCPO Box 2631, 0718 Makati
Metro Manila, Philippines
Contact No.(s) : (Tel.) 8189283, 8180466
(Fax) 8163183
Subject of Interest
Major : Engineering, integrated farming systems, rice-fish farming systems
Minor : Brackishwater aquaculture, freshwater aquaculture, tilapias, training, water quality

107. Mr. Inza **Dembele**
Department pisciole, Institut des
Savanes, 01BP 633, Bouake 01,
Cote d'Ivoire
Contact No.(s) : (Tel.) 632170
(Fax) 632045
Subject of Interest
Major : Farming systems research, production systems, tilapias

108. Mr Hamet D. **Diadhiou**
Centre de recherches oceanographiques
de Dakar-Thiaroye, BP 427 Ziquinchor,
Senegal
Contact No.(s) : (Tel.) 340534
Subject of Interest
Major : Bivalves, brackishwater aquaculture
Minor : Growth

109. Mr. Anis **Diallo**
Aquaculture, Centre de recherches
oceanographiques de Dakar-Thiaroye
BP 427 Ziquinchor, Senegal

Contact No.(s) : (Tel.) 340534 to 36
(Fax) 342792

Subject of Interest

Major : Brackishwater aquaculture, environmental research, farming management, farming systems, pond management, production systems, rich-fish farming systems, tilapias, waste-fed, water quality, zooplankton

Minor : Computer methods, economics, freshwater aquaculture, growth, marketing, resources economics, rural sociology, socioeconomics

110. Dr. Malcolm W. **Dickson**
Project Company Manager
Central and Northern Regions Fish
Farming Project PO Box 700,
Mzuzu, Malawi
Contact No.(s) : (Tel.) 335253
(Fax) 335051

Subject of Interest

Major : Integrated farming systems, nutrition, tilapias

Minor : Development, farming management, freshwater aquaculture

111. Mr. Khokal R. **Dinesh**
Research Scholar, Genetics Laboratory
Department of Zoology
National University of Singapore
Lower Kent Ridge, Singapore 0511
Contact No.(s) : (Tel.) 7722707
(Fax) 7792486

Subject of Interest

Major : DNA fingerprinting, genetics

Minor : Carps, genetic resources, induced spawning

112. Mr. Nelson M. **Doren**
Assistant Professor II, School of Fisheries,
Mariano Marcos State University,
Currimao, Ilocos Norte 2903, Philippines

Subject of Interest
Major : Extension, freshwater aquaculture, integrated farming systems

113. Dr. Thierry **Dodet**
SEPIA International,
14 av. Custave Eiffel
78182 Saint-Quentin en Yvelines Cedex,
France
Contact No.(s) : (Tel.) 30964975
(Fax) 30964949
Subject of Interest
Major : Environmental studies, nutrition
Minor : Brackishwater aquaculture, tilapias

114. Dr. Kiran **Dube**
Scientist/Head of Section
Central Institute of Fisheries
Education, Versova,
Bombay 400061
India
Contact No.(s) : (Tel.) 6261446 to 48
(Fax) 6261573
Subject of Interest
Major : Integrated farming systems, pond dynamics, reproductive physiology
Minor : Limnology, production systems, rich-fish farming systems

115. Dr. Henri J. **Dumont**
Professor,
Institute of Animal Ecology
University of Ghent
Ledeganckstraat, 35, Ghent 9000, Belgium
Contact No.(s) : (Tel. / Fax) 645253
Subject of Interest
Major : Freshwater aquaculture, zooplankton
Minor : Crustaceans, ecology limnology

116. Mr. Danilo T. **Dy**
Research / Administrative Officer
Marine Biology Section
University of San Carlos, Cebu City
6000, Philippines
Subject of Interest
Major : Ecophysiology, environmental impact, meiofauna
Minor : Computer methods, software, water quality

117. Dr. Joseph Effiong Eyo
Fisheries & Hydrobiology Research Unit,
Department of Zoology
PO Box 3146, University of Nigeria,
Nsykla enugu state, Nigeria
Contact No.(s) : (Tel.) 042-771911
(Fax) 042-770644, 771500
Subject of Interest
Major : Cat fish breeding, Genetics and Taxonomy
Minor : Aquaculture, nutrition
Subject of Consultancy
Fish breeding, genetics and nutrition.

118. Dr. Udeni **Edirisinghe**
Senior Lecturer, Department of Animal
Science, Faculty of Agriculture,
University of Peradeniya
Peradeniya, Sri Lanka
Contact No.(s) : (Tel.) 88657, 88239
(Fax) 88041
Subject of Interest
Major : Freshwater aquaculture, integrated farming systems, reservoirs

119. Dr. Peter **Edwards**
Professor
Division of Agricultural and Food
Engineering, AIT, GPO Box 2754,
Bangkok 10501, Thailand
Contact No.(s) : (Tel.) 5160110
(Fax) 5162126

Subject of Interest
Major : Integrated farming systems, pond dynamics, wastewater (sewage) reuse
Minor : Carps, tilapias, wastefed

120. Mr. Sunday Bl **Ekanem**
Lecturer I, Department of Biological Sciences, University of Calabar, PMB 1115, Calabar, Nigeria
Contact No.(s) : (Tel.) 222749
(Fax) 221773, 221137
Subject of Interest
Major : Biochemistry, cat-fishes, feed technology, finfish, freshwater aquaculture.
Minor : Growth, induced spawning, information, nutrition, water quality

121. Dr. Santiago U.K. **Ekaratne**
Senior Lecturer Grade 1,
Department of Zoology,
University of Colombo,
PO Box 1490, Colombo 3, Sri Lanka
Conact No.(s) : (Tel.) 503399
Subject of Interest
Major : Bivalves, coral reefs, crustaceans, lagoons
Minor : Growth, physiology

122. Dr. Ambekar E. **Eknath**
Associate Scientist, ICLARM
MCOP Box 2631, 0718 Makati
Metro Manila, Philippines
Contact No.(s) : (Tel.) 8189283, 8180466, 8175255
(Fax) 8163183
Subject of Interest
Major : Genetic resources, genetics, modeling
Minor : Farming systems research, population genetics

123. Mr. Ekpo G. **Ekpo**
Senior Research Technologist
Department of Fisheries and Aquaculture
Institute of Oceanography,
University of Calabar,
PMB 1115, Calabar, Nigeria

Contact No.(s) : (Tel.) 222749
(Fax) 221773, 221137
Subject of Interest
Major : Freshwater aquaculture, hatchery technology, induced spawning
Minor : Diseases, engineering nutrition

124. Mr. Michael O. **Ella**
Deputy Director of Fisheries,
Department of Fisheries, PMB 1168, Calabar,
Cross River State, Nigeria
Subject of Interest
Major : Catfishes, extension, feed technology
Minor : Endocrinology, integrated farming systems, pond management

125. Mr Abdalla G. **Elnaiem**
Associate Professor,
Red Sea Fisheries Research Centre,
Box 730, Port Sudan,
Sudan
Contact No.(s) : (Tel.) 4667
Subject of Interest
Major : Bivalves, farming management, marine aquaculture

126. Mr. Udeme I **Enin**
Research Fellow II, Institute of Oceanography,
University of Calabar, PMB 1115,
Calabar, Nigeria
Contact No.(s) : (Tel.) 222749
(Fax) 221773, 221137
Subject of Interest
Major : Curriculum development, farming management, post-harvest technology
Minor : Conservation, resource economics

127. Ms. Mamaa **Entsua**-Mensah
Research Officer,
Institute of Aquatic Biology,
PO Box 38, Achimota, Ghana

Contact No.(s) : (Tel.) 775511 (Accra)
(Fax) 777655 (Accra)
Subject of Interest
Major : Software, statistics, stock assessment

128. Dr. John M. **Epifanio**
Research Associate, Department of Biology
Virginia Commonwealth University,
Box 2012, 816 Park Ave., Richmond,
VA 23284-2012, USA
Contact No.(s) : (Tel.) 3671562
(Fax) 3670503
Subject of Interest
Major : Finfish, genetic resources
Minor : Policy

129. Ms. Cornelia C. **Estabillo**
Assistant Professor,
College of Fisheries, Pangasinan State University,
Binmaley
Panagsinan 2417, Philippines
Subject of Interest
Major : Freshwater aquaculture, seed production tilapias
Minor : Giant clams, reservoirs

130. Mr. Lawrence E. **Etim**
Research Fellow / Lecturer
Department of Fisheries and Aquaculture
Institute of Oceanography,
University of Calabar,
PMB 1115, Calabar, Nigeria
Contact No.(s) : (Tel.) 222749
(Fax) 221773, 221137
Subject of Interest
Major : Bivalves, finfish, stock assessment

131. Mr. Joseph E. **Eyo**
Lecturer, Department of Zoology
University of Nigeria, Enugu State
Nigeria

Subject of Interest
Major : Catfishes, freshwater aquaculture, seed production
Minor : Genetics, reproductive physiology, taxonomy

132. Dr. Bernard **Ezenwa**
Chief Research Officer
Nigerian Institute for Oceanography
and Marine Research
PMB 12729, Victoria Island, Lagos, Nigeria
Contact No.(s) : (Tel.) 617530
Subject of Interest
Major : Crustaceans, hatchery technology, induced spawning
Minor : Economics, farming management, genetics

133. Mr. Ulunga **Faanunu**
Fisheries Officer, Ministry of Fisheries
PO Box 871, Nukualofa, Tonga
Contact No.(s) : (Tel.) 21399
(Fax) 23891
Subject of Interest
Major : Marine aquaculture
Minor : Resource economics

134. Dr. Oyedapo A. **Fagbenro**
Ph.D. Student, Institute of Aquaculture
University of Stirling, Stirling
FK9 4LA, UK
Contact No.(s) : (Fax) 472133
Subject of Interest
Major : Finfish, nutrition
Minor : Ecology

135. Mr. Jerry O. **Fagolimul**
Marine Research Specialist
Marine Resources Management Division
Department of Resources and Development
PO Box 251, Yap FM 96943, Micronesia
Contact No.(s) : (Tel.) 2294

Subject of Interest
Major : Giant clams, hatchery technology, seed production
Minor : Marine aquaculture, site selection, stock assessment

136. Ms. Emma **Fajer Avila**
Chief of Department
Centro de Investigaciones Pesgueras
5th Ave y 248 Edificio 2, Barloventro
Playa Ciudad Havana, Cuba
Contact No.(s) : (Tel.) 227645
Subject of Interest
Major : Disease, immunology, pathology
Minor : Crustaceans, freshwater aquaculture, microbiology

137. Dr. Augustine E. **Falaye**
Department of Wildlife & Fisheries Management,
University of Ibadan,
Ibadan, Nigeria
Contact No.(s) : (Tel.) 400550 to 614
Subject of Interest
Major : Nutrition
Minor : Freshwater aquaculture, tilapias

138. Dr. Anthony S.D. **Farmer**
Anthony Farmer & Associates, Prime Farm
Whitchurch Canonicorum Bridport
Dorest DT6 6RP, UK
Contact No.(s) : (Tel.) 89100, 89736
(Fax) 89081
Subject of Interest
Major : Brackishwater aquaculture, crustaceans, marine aquaculture
Minor : Environmental impact, rich-fish farming systems, shrimps

139. Mr. Daniel F. **Fegan**
Research Manager, Aquastar Laboratories
PO Box 5, Amphur Muang, Songkhla 90000,
Thailand
Contact No.(s) : (Tel.) 321056
(Fax) 321057

Subject of Interest
Major : Hatchery technology, shrimps
Minor : Diseases, nutrition, water quality

140. Mr. S. **Felix**
Assistant Professor, Department of
Aquaculture, Fisheries College and Research
Institute, Tuticorin 628008, Tamil Nadu,
India
Contact No.(s) : (Tel.) 22154, 21554
Subject of Interest
Major : Brackishwater aquaculture, pathology
Minor : Conservation, integrated farming systems

141. Ms. Becky A. **Field**
Network Administrator, Mollusc Culture
Network, Biology Department, Dalhousie
University, Halifax
Nova Scotia, Canada B3H 4J1
Contact No.(s) : (Tel.) 4943610
(Fax) 4946899
Subject of Interest
Major : Bivalves, marine aquaculture
Minor : Rural sociology

142. Dr. Lev **Fishelson**
Professor, Department of Zoology,
Tel Aviv University 69978
Tel Aviv, Israel
Contact No.(s) : (Tel.) 6408655
(Fax) 6409403
Subject of Interest
Major : Ecology, freshwater aquaculture
Minor : Behavior, reproductive physiology

143. Dr. Romeo D. **Fortes**
Professor, Institute of Aquaculture
College of Fisheries
University of the Phlippines in the
Visayas, Miang-ao, Iloilo 5023,
Philippines

Contact No.(s) : (Tel.) 81535
(Fax) 79248

Subject of Interest
Major : Brackishwater aquaculture, pond dynamics, pond management
Minor : Bivalves, marine aquaculture

144. Dr. Emmanuel **Frempong**
Senior Lecturer, Department of Biological Science, University of Science and Technology, Kumasi, Ghana
Contact No.(s) : (Tel.) 5350

Subject of Interest
Major : Freshwater aquaculture, integrated systems
Minor : Limnology

145. Dr. Rainer **Froese**
Associate Scientist, ICLARM
MCPO Box 2631, Makati
Metro Manila, Philippines
Contact No.(s) : (Tel.) 8189283, 8180466, 8175255
(Fax) 8163183

Subject of Interest
Major : Computer methods, conservation, information
Minor : Ecology, finfish, physiology

146. Dr. Jacques **Fuchs**
Institute francais de recherche pour l'expoloitation de la mer
155, rue Jean-Jacques Rousseau
92138 Issy-les-Moulineaux Cedex, France
Contact No.(s) : (Tel.) 46482260
(Fax) 46482276

Subject of Interest
Major : Finfish, marine aquaculture, socioeconomics
Minor : Environmental studies, nutrition, site selection

147. De. Gonzalo M. **Gajardo**
Lecturer, Institute Professional de Osorno
PO Box 933, Osorno, Chile
Contact No.(s) : (Tel.) 235377 L. 63
(Fax) 239517

Subject of Interest
Major : Genetics, marine aquaculture
Minor : Environmental impact

148. Dr. Graham A.E. **Gall**
Professor, Department of Animal Science
University of California, Davis, CA 95616
USA
Contact No.(s) : (Tel.) 7521257
(Fax) 7521254
Subject of Interest
Major : Broodstock, finfish, quantitative genetics
Minor : Conservation, genetic resources, population genetics

149. Dr. Ofelia **Galman-Omitogun**
Associate Professor, Institute of Biology
College of Science, University of the
Philippines, Diliman 1101, Quezon City,
Philippines
Contact No.(s) : (Tel.) 995071, 982471 L. 6452
(Fax) 996868
Subject of Interest
Major : Genetics, tilapias
Minor : Biotechnology, freshwater aquaculture

150. Mr. Ajit J. **Galvankar**
Fisheries Biologist, Department of
Fisheries, College of Agriculture
111/1, B. Prabhat Rd. Lane 14
Pune 411004, India
Contact No.(s) : (Tel.) 334527
Subject of Interest
Major : Brackishwater aquaculture, freshwater aquaculture nutrition
Minor : Seed production, site selection, tilapias

151. Mr. Luis Ma. B. **Garcia**
Scientist, SEAFDEC Aquaculture Department
Tigbauan, Iloilo 5021, Philippines

Contact No.(s) : (Tel.) 81261, 9230201 to 3 (Manila)
(Fax) 81340

Subject of Interest

Major : Endocrinology, induced spawning, reproductive physiology

Minor : Broodstock, milkfish, physiology

152. Mr. Ger **Guo Chang**
Associate Professor, Ocean University of Qingdao, 266003 Qingdao, China
Contact No.(s) : (Tel.) 264361
(Fax) 279091

Subject of Interest

Major : Cages, finfish, genetics

Minor : Brackishwater aquaculture, ecology, growth

153. Mr. Mark H. **Gervis**
Manager, ICLARM Coastal Aquaculture Centre, PO Box 438, Honiara, Solomon Islands
Contact No.(s) : (Tel.) 20255
(Fax) 22130

Subject of Interest

Major : Bivalves, giant clams, marine aquaculture

Minor : Coral reefs, farming management, production systems

154. Dr. Teferra **Getachew**
Associate Professor, Department of Biology, Addis Ababa University, PO Box 1176 Addis Ababa, Ethiopia
Contact No.(s) : (Tel.) 553177 L. 152
(Fax) 550911

Subject of Interest

Major : Nutrition, physiology tilapias

Minor : Freshwater aquaculture, growth, limnology

155. Dr. Adel N.Z. **Gindy**
Head of Aquaculture Section
Marine Science and Fisheries Centre
Ministry of Agriculture and Fisheries
Resources, Sultanate of Oman,
PO Box 467, Muscat, Egypt
Contact No.(s) : (Tel.) 740061
(Fax) 740159
Subject of Interest
Major : Cages, marine aquaculture, rich-fish farming systems
Minor : Bivalves

156. Dr. Trygve **Gjedram**
Professor,
Institute of Aquaculture Research,
Agriculture University of Norway,
PO Box 10, N-1432 AS, Norway
Contact No.(s) : (Tel.) 949513
(Fax) 949502
Subject of Interest
Major : Quantitative genetics
Minor : Freshwater aquaculture, marine aquaculture

157. Dr. Bjarne R. **Gjerde**
Scientist,
Institute of Aquaculture Research,
Agricultural University of Norway
PO Box 10, 1432 AS-NLH,
Norway
Contact No.(s) : (Tel.) 949514
(Fax) 949502
Subject of Interest
Major : Quantitative genetics
Minor : Carps, tilapias

158. Mr. Hans M. **Gjoen**
Institute of Aquaculture Research
Agricultural University of Norway
PO Box 10, 1432 As-NLH, Norway

Contact No.(s) : (Tel.) 942515
(Fax) 949502
Subject of Interest
Major : Quantitative genetics
Minor : Marine aquaculture, tilapias

159. Mr. Israil Md. **Golder**
Farm Manager, M.I. Golder
Fish Hatchery Development Project
Parbatipur 5250, Dinajpur, Bangladesh
Subject of Interest
Major : Environmental studies, freshwater aquaculture integrated farming systems
Minor : Extension, nutrition, pesticides

160. Dr. Edgardo **D. Gomez**
Professor/Director, Marine Science
Institute, University of the Philippines,
Diliman 1101 Quezon City,
Philippines
Contact No.(s) : (Tel.) 989676 to 9 L. 7416
(Fax) 974355
Subject of Interest
Major : Giant clams, habitat restoration, marine aquaculture
Minor : Coral reefs, seed production, site selection

161. Dr. Julian F. **Gonsalves**
Vice President for Programs, IIRR
Silang, Cavite 4118, Philippines
Contact No.(s) : (Tel.) 582659 (Manila)
(Fax) 5222494 (Manila)
Subject of Interest
Major : Conservation, extension, integrated farming systems
Minor : Habitat restoration, pond management, rice-fish farming systems

162. Mr. Dhaneshwar **Goorah**
Divisional Scientific Officer
Albion Fisheries Research Centre,
Albion, Mauritius
Contact No.(s) : (Tel.) 2334729
Subject of Interest
Major : Crustaceans, diseases, stock assessment

163. Dr. Moshe **Gophen**
Senior Scientist/Full Professor
Biological Station, University of
Oklahoma, HC 71, Box 205, Kingston,
OK 73479, USA
Contact No.(s) : (Tel.) 5642463
(Fax) 5642479
Subject of Interest
Major : Limnology, tilapias, zooplankton
Minor : Ecophysiology, food-chains, water quality

164. Mr. Christopher **Gordon**
Research Fellow, Vo ta Basin Research
Project, University of Ghana,
PO Box 209, Legon, Accra, Ghana
Subject of Interest
Major : Brackishwater aquaculture, crustaceans, environmental studies
Minor : Reservoirs, water quality, zooplankton

165. Dr. Philippe **Goulletquer**
Researcher, Aquaculture Research
Laboratory, Institut Francais de
recherche pour l'exploitation de
la mer Avenue du General de Gaulle,
BP 32 14520 Port-en-Bessin-Huppain,
France
Contact No.(s) : (Tel.) 31511300
(Fax) 31511301
Subject of Interest
Major : Bivalves, ecophysiology, marine aquaculture
Minor : Economics, growth, modeling

166. Mr. Hugh **Govan**
Servicenter, Terraza 4 Estepone
29680, Malaga, Spain
Contact No.(s) : (Tel.) 895064
(Fax) 803386
Subject of Interest
Major : Bivalves, coral, giant clams
Minor : Extension, marine aquaculture, predators

167. Mr. **Gu Ruobo**
Asian Pacific Regional Research
and Training Centre for Integrated
Fish Farming, Wuxi, China 214081
Contact No.(s) : (Tel.) 601424
(Fax) 603304
Subject of Interest
Major : Freshwater aquaculture, induced spawning, integrated systems
Minor : Hatchery technology, pond management, rice-fish farming systems

168. Dr. Rafael D. **Guerrero III**
C/o Aquatic Biosystems, Bay, Laguna 4300,
Philippines
Subject of Interest
Major : Freshwater aquaculture, marine aquaculture, tilapias
Minor : Farming management, hatchery technology, training

169. Ms. Rasanthi M. **Gunasekera**
Ph.D. Student, Department of Zoology
National University of Singapore
Lower Kent Ridge, Singapore 0511
Contact No.(s) : (Tel.) 7722708
(Fax) 7792486
Subject of Interest
Major : Nutrition
Minor : Reproductive physiology, tilapias

170. Dr. Jorge N. **Gunther**
Professor, Escuela Ciencias Biologicas
Universidad Nacional, Heredia,
Costa Rica
Contact No.(s) : (Tel.) 375230
(Fax) 376427
Subject of Interest
Major : Bioenergetics, finfish, nutrition

171. Dr. Modadugu V. **Gupta**
Senior Aquaculture Specialist
ICLARM Bangladesh Office
House 20, Rd. 9/A, Dhanmondi R/A
Dhaka 1209, Bangladesh
Contact No.(s) : (Tel.) 812466, 817345
(Tlx) 642 671 HBP BJ
(Fax) 884968, 883416

Subject of Interest

Major : Farming systems research, freshwater aquaculture, production systems

172. Mr. Richard **Gurure**
Department of Animal Science
University of Guelph, Guelph
Ontario, Canada N1G 4S7

Subject of Interest

Major : Farming systems research, feed technology, finfish, integrated farming systems

173. Dr. Mohammad Gulam **Hussain**
Fresh water Station
Bangladesh Fisheries Research Institute
Mymensingh 2201, Bangladesh
Contact No.(s) : (Tel.) 880-91-54221
880-91-54375 (R)
(Fax) 880-91-55259, 880-2-883416

Subject of Interest

Major : Aquaculture, fish genetics and reproductive biology

Minor : Aquaculture biodiversity, documentation of threatened fish species.

Subject of Consultancy

Fish hatchery operation and management.
Fish seed production and nursery management.
Development of Fish breeding plans and stock improvement.
Development of monosex population through sex reversal technique of Tilapia and minor carp species.
Fish reproductive biology physiology
Aquaculture of fin fishes.
Integrated fish-livestock, agriculture farming.

174. Dr. Robert William **Hickran**
Aquaculture Research Centre,
NIWA, PO Box 14-901
Wellington, New Zealand
Contact No.(s) : (Tel.) +64-4-3888596
(Fax) +64-4-3889931

Subject of Interest
Major : Aquaculture/marine farming of mussels, oysters and flatfish
Minor : Temperate and tropical aquaculture of shell fish and fish commercial technology production etc.

Subject of Consultancy
Mussel farming development/assessment, oyster farming, development/assessment, flat fish farming techniques.

175. Dr. Kirk **Hahn**
Researcher, HIMB/Coconut Island
PO Box 1346, Kaneohe, HI 96744, USA

Subject of Interest
Major : Endocrinology, marine aquaculture, physiology

176. Mr. Boyd A. **Haight**
Senior Aquaculturist,
C/o FAO Representative, PO Box 3730,
Harare, Zimbabwe
Contact No.(s) : (Tel.) 724985
(Fax) 729563

Subject of Interest
Major : Extension, freshwater aquaculture
Minor : Computer methods, environmental impact, reservoirs

177. Mr. Rene D. **Haller**
Managing Director, Baobab Farm Ltd.
PO Box 81995, Mombasa, Kenya
Contact No.(s) : (Tel.) 486155, 485754, 485729, 485501
(Fax) 485151

Subject of Interest
Major : Freshwater aquaculture, tilapias
Minor : Rice-fish farming systems, wastewater (sewage) reuse

178. Dr. Eric M. **Hallerman**
Assistant Professor
Department of Fisheries & Wildlife Sciences
Virginia Polytechnic Institute and State
University Blacksburg, VA 24061-0321, USA
Contact No.(s) : (Tel.) 2313257
(Fax) 2317580
Subject of Interest
Major : Genetics, policy, population genetics
Minor : Biotechnology, conservation, curriculum development

179. Mr. Matthias **Halwart**
Scientific Employee, University of Stuttgart,
University of Landscape Planning &
Ecology Keplerstr. II, 7000 Stuttgart I
Subject of Interest
Major : Ecology, freshwater aquaculture, rice-fish farming systems
Minor : Integrated pest management

180. Mr. A.K. Yousuf **Haroon**
Senior Scientist, Fisheries Research
Institute Riverine Station, Baburhat
Chandpur 3602, Bangladesh
Contact No.(s) : (Tel.) 3482, 3407
Subject of Interest
Major : Freshwater aquaculture, hatchery technology, limnology
Minor : Pond management, rice-fish farming systems, stock assessment

181. Dr. Sheenan **Harpaz**
Aquaculture Research Specialist
Fish & Aquaculture Research Unit
Agricultural Research Organization
PO Box 6, Bet Dagan, Israel
Contact No.(s) : (Tel.) 9683388
(Fax) 993998

Subject of Interest
Major : Behaviour, freshwater aquaculture, nutrition
Minor : Crustaceans, growth, integrated farming systems

182. Dr. Brian J. **Harvey**
President, MTL Biotech Ltd.,
PO Box 5760, Station B,
Victoria, B.C., Canada V8R 6S8
Contact No.(s) : (Tel.) 5986019
(Fax) 5982047
Subject of Interest
Major : Biotechnology, cryopreservation, genetic resources
Minor : Bivalves, environmental studies

183. Dr. Mohammad R. **Hasan**
Associate Professor
Department of Fisheries Biology
and Limnology, Bangladesh Agricultural
University, Mymensingh 2202, Bangladesh
Contact No.(s) : (Tel.) 5695 to 97 L. 281
Subject of Interest
Major : Carps, feed technology, nutrition
Minor : Biotechnology, environmental impact, stock assessment

184. Dr. Roshada **Hashim**
Lecturer, School of Biological Sciences
University Sains Malaysia
11800 Minden, Penang, Malaysia
Contact No.(s) : (Tel.) 877888 L. 3048
(Fax) 871526
(Tlx) USM MA 40254
Subject of Interest
Major : Extension, nutrition
Minor : Finfish

185. Mr. Graham S. **Haylor**
Research Fellow / Consultant, Institute
of Aquaculture, University of Stirling,
Stirling FK9 4LA, UK

Contact No.(s) : (Tel.) 473171
(Fax) 472133

Subject of Interest

Major : Catfishes, farming management, hatchery technology

Minor : Farming systems research, production systems, training

186. Dr. Dennis **Hedgecock**
Professor, Bodega Marine Laboratory
PO Box 247, Bodega Bay, CA 94923, USA
Contact No.(s) : (Tel.) 8752075
(Fax) 8752089

Subject of Interest

Major : Genetic resources, population genetics

Minor : Bivalves, crustaceans, finfish

187. Mr. Robert W. **Hickman**
Aquaculture Scientist, MAF Fisheries
Aquaculture Research Centre, PO Box 297
Wellington, New Zealand
Contact No.(s) : (Tel.) 3888596
(Fax) 3863179

Subject of Interest

Major : Bivalves, farming systems research, marine aquaculture

Minor : Hatchery technology, induced spawning, seed production

188. Dr. Kevin D. **Hopkins**
Associate Professor, College of Agriculture,
University of Hawaii at Hilo,
96720-4091, USA
Contact No.(s) : (Tel.) 9333393
(Fax) 9333674

Subject of Interest

Major : Curriculum development, finfish, pond dynamics

Minor : Growth, integrated farming systems, tilapias

189. Dr. Bram E.A. **Huisman**
Professor, Department of Fish Culture
and Fisheries, Wageningen Agricultural
University PO Box 338, 6700 AH Wageningen,
The Netherlands
Contact No.(s) : (Tel.) 83307
(Fax) 83962
Subject of Interest
Major : Finfish, freshwater aquaculture, growth
Minor : Bioenergetics, carps, catfishes, farming management

190. Dr. Gideon **Hulata**
Research Scientist, Fish & Aquaculture
Research Unit, Agriculture Research
Organization, Volcani Centre
PO Box 6, Bet Dagan 50250, Israel
Contact No.(s) : (Tel.) 9683388
(Fax) 993998
Subject of Interest
Major : Carps, genetics, tilapias
Minor : Biotechnology, cryopreservation

191. Dr. Anicia Q. **Hurtado-Ponce**
Scientist 1, SEAFDEC Aquaculture
Department, Tigbauan, Iloilo 5021,
Philippines
Contact No.(s) : (Tel.) 81261, 9230201 to 3 (Manila)
(Fax) 81340
Subject of Interest
Major : Seaweeds
Minor : Brackishwater aquaculture, marine aquaculture

192. Mr. M.G. **Hussain**
Chief Scientific Officer, Fisheries
Research Institute Mymensingh,
Bangladesh
Subject of Interest
Major : Biotechnology, genetics, hatchery technology
Minor : Endocrinology, farming management, freshwater aquaculture

193. Mr. Johannes **Hutabarat**
Lecturer, Department of Aquaculture
and Fisheries, Diponegoro University
J1. Hayamwuruk No. 4A Semarang, Indonesia
Contact No.(s) : (Tel.) 516029
(Fax) 218381
Subject of Interest
Major : Brackishwater aquaculture, crustaceans, feed technology
Minor : Environmental impact, finfish, hatchery technology, marine aquaculture

194. Mr. Jorgen **Hylleberg**
Associate Professor, Institute of
Biological Sciences, Department of
Genetics and Ecology, Arhus University
Ny Munkegade, 8000 Arhus C, Denmark
Contact No.(s) : (Tel.) 127188 L. 5157
(Fax) 127191
Subject of Interest
Major : Coral reefs, detritus, marine aquaculture

195. Mr. Adekunle, **A. Idowu**
Senior Biology Teacher
Anglican Girls' Grammar School
PO Box 35, Ijebu-Ode, Ogun State,
Nigeria
Subject of Interest
Major : Freshwater aquaculture, nutrition, parasities
Minor : Ecology, zooplankton

196. Dr. John E.G. **Ikeorgu**
Coordinator, Farming Systems Program
National Root Crops Research Institute
Umudike, PMB 7006, Umuahia, Abia State,
Nigeria
Subject of Interest
Major : Farming systems research, integrated farming systems, rich-fish farming systems
Minor : Freshwater aquaculture, water-feed

197. Mr. Sudhakar T. **Indulkar**
Biologist
Taraporevala Marine Biological Research
Station, Taraporevala Aquarium, N.S. Road,
Bombay 400002, India
Contact No.(s) : (Tel.) 313333, 312061
Subject of Interest
Major : Brackishwater aquaculture, feed technology freshwater aquaculture
Minor : Hatchery technology, pond management, seed production.

198. Mr. **Iskandar**
Lecturer, Department of Fisheries
Padjadjaran University
J1, Raya Bandung-Sumedang, Km. 21
Jatinangor 45363, Sumedang, West Java,
Indonesia
Contact No.(s) : (Tel.) 443998
Subject of Interest
Major : Freshwater aquaculture, seed production
Minor : Environmental studies

199. Ms. Sri **Ismawati**
Researcher
Balai Penelitian Perikanan Budidaya
Pantai J1. Makmur Dg, Sitakka, Maros 90511
Sulawesi, Selatan, Indonesia
Contact No.(s) : (Tel.) 24940
Subject of Interest
Major : Bivalves, Brackishwater aquaculture, water quality
Minor : Biotechnology, computer methods, engineering, reproductive physiology, site selection, software, statistics, stock assessment

200. Dr. Andrej **Ivoilov**
Laboratory of Ichthyology
Biological Institute of St. Petersburg
Stary Peterhof, Oranienbaumskoye Sch. 2
198904, St. Petersburg, USSR

Contact No.(s) : (Tel.) 1068839
Subject of Interest
Major : Reproductive physiology, tilapias, water systems
Minor : Freshwater aquaculture, taxonomy

201. Mr. Daniel M. **Jamu**
Research Associate
ICLARM/GTZ Africa Aquaculture Project
PO Box 229, Zomba, Malawi
Contact No.(s) : (Tel.) 531274, 531215
(Tlx) 4441 BUREAU MI
(Fax) 528103
Subject of Interest
Major : Pond dynamics
Minor : Environmental studies, integrated farming systems

202. Ms. Carmela **C. Janagap**
Research Assistant, ICLARM
MCPO Box 2631, 0718 Makati
Metro Manila, Philippines
Contact No.(s) : (Tel.) 8189283, 8180466, 8175255
(Fax) 8163183
Subject of Interest
Major : Computer methods, modeling, statistics
Minor : Genetics, software, tilapias, training

203. Mr. Johannes A.L. **Janssen**
Project Manager
BP 76, Antsirabe, Madagascar
Contact No.(s) : (Fax) 49709
(Tlx) 22516 FAO MG
Subject of Interest
Major : Extension, pond management, rice-fish farming systems
Minor : Carps, nutrition, reproductive physiology, tilapias

204. Ms. Parnsri **Jarimopas**
National Inland Fisheries Institute
Kasetsart University Campus
Bangkok 10900, Thailand
Contact No.(s) : 5794121

Subject of Interest
Major : Catfishes, seed production, tilapias
Minor : Cages, genetics

205. Dr. Kim **Jauncey**
Lecturer, Institute of Aquaculture
University of Stirling, Stirling
FK9 4LA, UK
Contact No.(s) : (Tel.) 467892
(Fax) 472133
Subject of Interest
Major : Biochemistry, feed technology, nutrition
Minor : Bioenergetics, software, tilapias

206. Dr. V. **Jayaprakas**
Reader, Department of Aquatic Biology
and Fisheries, University of Kerala,
Beach PO Box 1603 Trivandrum--Pin 695007
Kerala, India
Contact No.(s) : (Tel.) 71138
Subject of Interest
Major : Biotechnology, brackishwater aquaculture, freshwater aquaculture
Minor : Cryopreservation, genetics, nutrition

207. Ms. Annesty P.M. **Jayasuriya**
Research Officer, National Aquatic
Resources Agency Crow Island, Mattakuliya,
Colombo 15, Sri Lanka
Contact No.(s) : (Tel.) 522000
Subject of Interest
Major : Coral reefs, seaweeds
Minor : Environmental studies

208. Dr. Yetunde C. **Jeje**
Lecturer, Fisheries and Hydrobiology Unit
Department of Zoology, University of Ibadan
Ibadan, Nigeria
Contact No.(s) : (Tel.) 400550 L. 2082

Subject of Interest
Major : Farming systems research, pond dynamics zooplankton

209. Mr. Andreas **Jensen**
FAO/Fisheries Department
PO Box 710005, Manza, Zambia
Subject of Interest
Major : Coral reefs, integrated farming systems, marine aquaculture
Minor : Environmental studies, extension, lagoons

210. Ms. Narriman S. **Jiddawi**
Research Fellow, Institute of Marine
Science Box 668, Zanzibar, Tanzania
Contact No.(s) : (Tel.) 30741, 32184
(Fax) 32128
Subject of Interest
Major : Bivalves, finfish, stock assessment
Minor : Coral reefs, marine aquaculture, seaweeds

211. Mr. Bonny B. **Jinivone**
Senior Fisheries Officer, Fisheries
Department, 8th Floor, Menara Khidmat
88628 Kota Kinabalu, Sabah, Malaysia
Contact No.(s) : (Tel.) 52641, 52541
(Fax) 240511
Subject of Interest
Major : Freshwater aquaculture, seed production
Minor : Cages, integrated farming systems

212. Dr. Alick R. **Jones**
Senior Lecturer/Department Head
Department of Pure and Applied Zoology
University of Reading, Whiteknights
PO Box 8, Reading RG6 2AJ, UK
Subject of Interest
Major : Pollution, water quality
Minor : Freshwater aquaculture

213. Mr. Clive M. **Jones**
Senior Biologist, Department of Primary
Industries, Fisheries Research Centre
Walkamin, Queensland 4872, Australia
Contact No.(s) : (Tel.) 933733
(Fax) 933903
Subject of Interest
Major : Crustaceans, fresh water aquaculture, pond management
Minor : Detritus, nutrition, production systems

214. Mr. Ronald W. **Jones**
57 Victoria St., Ottawa, Ontario, Canada
K1M 1S8
Contact No.(s) : (Tel.) 7456776
Subject of Interest
Major : Endocrinology, genetic resources, induced spawning, integrated farming systems, pesticides, reproductive physiology, rice-fish farming systems, seaweeds
Minor : Brackishwater aquaculture, broodstock, ecology, environmental studies, freshwater aquaculture

215. Dr. N. **Kaliaperumal**
Sr. Scientist
Regional Centre of Central Marine
Fisheries Research Institute
Marine Fisheries-623520
Ramnad District, Tamil Nadu, India
Contact No.(s) : (Tel.) 04573-41443, 41456
08572-27666 (R)
(Fax) 04573-41502
Subject of Interest
Major : Sea weeds Marine Algae (ecology, systematics, biology, chemistry, resources, culture and utilisation)
Minor : Sea grasses (ecology, systematics resources, biology and utilisation)

Subject of Consultancy

Cultivation of agar yielding seaweeds in the sea and onshore tanks.

Project preparation erection of agar unit and production of agar (china grass)

Project preparation, erection of algin unit and production of alginates.

Survey of seaweeds and sea grasses in shallow and deep waters of the sea.

216. Mr. Winston K. **Kadongola**
Scientific Officer
Department of Research and Environmental
Affairs, PO Box 30745, Lilongwe 3, Malawi
Contact No.(s) : (Tel.) 781111
(Fax) 781487

Subject of Interest

Major : Environmental impact, environmental studies freshwater aquaculture

217. Mr. Edouard **Kali-Tchikati**
Chef de Service
Project pisciculture rurale de Djoumduna
BP 675, Brazzaville, Congo
E-mail : BP 98

Subject of Interest

Major : Extension, freshwater aquaculture, integrated farming systems, tilapias

218. Dr. Jim M. **Kapetsky**
Senior Fishery Resources Officer, FAO
Inland Waters and Aquaculture Service, FIRI
Rm. F-231, Via delle Terme di Caracalla
00100 Rome, Italy
Contact No.(s) : (Tel.) 57976646
(Fax) 5120330

Subject of Interest

Major : Development, policy, site selection

219. Dr. Andalwisye M. **Katule**
Senior Lecturer,
Sokoine University of Agriculture,
PO Box 3004, Chuo Kikuu,
Morogoro, Tanzania
Contact No.(s) : (Tel.) 1515
Subject of Interest
Major : Bivalves, hatchery technology, marine aquaculture

220. Mr. Emmanuel K.W. **Kaunda**
Research Assistant
ICLARM/GTZ Africa Aquaculture Project
PO Box 229, Zomba, Malawi
Contact No.(s) : (Tel.) 531274, 531215
(Fax) 528103
Subject of Interest
Major : Modeling, stock assessment
Minor : Statistics

221. Mr. Norbert A. **Kayombo**
Ph.D. Candidate,
Humboldt University
National History Museum
Invalidenstrasse 43, 0-1040 Berlin,
Germany
Contact No.(s) : (Tel.) 98972665
Subject of Interest
Major : Bivalves, freshwater aquaculture, marine aquaculture

222. Dr. Clive P. **Keenan**
Fisheries Biologist,
Southern Fisheries Centre,
PO Box 76, Deception Bay,
Queensland 4508, Australia
Contact No.(s) : (Tel.) 2031444
(Fax) 2033517
Subject of Interest
Major : Conservation, population genetics, taxonomy
Minor : Freshwater aquaculture, genetic resources, marine aquaculture

223. Mr. Romanus I. **Keke**
Lecturer, School of Biological Sciences
Abia State University, Uturu
PMB 2000, Okigwe, Nigeria
Subject of Interest
Major : Induced spawning, nutrition, seed production
Minor : Broodstock, freshwater aquaculture

224. Dr. Craig B. **Kensler**
International Consultant
44 Vale St., Denbigh, Clwyd LL 16 3AH, UK
Contact No.(s) : (Tel.) 914323
Subject of Interest
Major : Training
Minor : Brackishwater aquaculture, marine aquaculture

225. Dr. Lotfi F. **Khalil**
Assistant Director
International Institute of Parasitology
395a Hatfield Road, St. Albans, Herts,
AL4 0XU, UK
Contact No.(s) : (Tel.) 833151
(Fax) 868721
Subject of Interest
Major : Diseases, parasites
Minor : Public health

226. Mr. Golam **Kibria**
Flat-2, 6 King St., Coburg
Melbourne, VIC 3058, Australia
Contact No.(s) : (Tel.) 3836815
Subject of Interest
Major : Marine aquaculture, production systems, water quality
Minor : Carps, extension, nutrition

227. Mr. Raouf W. **Kilada**
Scientific Visitor, University of Texas
at Austin, Marine Science Institute,
PO Box 1267, Port Aransas
Texas 78373-1267, USA

Contact No.(s) : (Tel.) 7496795
(Fax) 7496749
Subject of Interest
Major : Coral reefs, ecology, giant clams
Minor : Bivalves, computer methods, environmental impact

228. Mr. Favognon J-P.E. **Kone**
Service Chief, 04 BP 283, Abidjan 04,
Cote d'Ivoire
Contact No.(s) : (Tel.) 301046
Subject of Interest
Major : Ecology, production systems, training
Minor : Conservation, diseases, environmental impact, limnology, public health

229. Mr. Hassanai **Kongkeo**
Technical Officer, ASEAN-EEC AADCP
PO Box 1006, Kasetsart Universtiy Campus
Bangkhen, Bangkok 10903, Thailand
Contact No.(s) : (Tel.) 5613022
(Fax) 5613023
Subject of Interest
Major : Brackishwater aquaculture, crustaceans, farming management
Minor : Diseases, marine aquaculture, pollution

230. Mr. N'Gouan Cyrille **Kouassi**
Department piscicole, Institut des Savanes
01 BP 633, Bouake 01, Cote D'Ivoire
Contact No.(s) : (Tel.) 632170
(Fax) 632045
Subject of Interest
Major : Freshwater aquaculture, zooplankton

231. Mr. Wayne L. **Kruckenberg**
Fish Biologist
Aquatic and Wildlife Resource Division
Department of Agriculture, PO Box 1771
Agana, Guam 96910, USA
Contact No.(s) : (Tel.) 4773311

Subject of Interest
Major : Conservation, finfish, pond management
Minor : Brackishwater aquaculture, freshwater aquaculture, water systems

232. Mr. Divine **Kuma**
Fisheries Officer, Tono Fish Farm
Irrigation Consultants of the Upper Regions
PO Box 780, Accra, Ghana
Contact No.(s) : (Tel.) 667021
Subject of Interest
Major : Freshwater aquaculture, integrated farming systems, rice-fish farming systems
Minor : Pond management, rural sociology, seed production

233. Dr. Arumugam K. **Kumaraguru**
Reader, School of Energy
Environment and Natural Resources
Madurai Kamaraj University
Madurai, Tamil Nadu 625021 India
Contact No.(s) : (Tel.) 85265-
Subject of Interest
Major : Bioenergetics, coral reefs, ecophysiology
Minor : Environmental studies, limnology, pollution

234. Dr. Andreas **Kunzmann**
Assistant Professor, Fisheries Faculty
Bung Hatta University, Ulak J1. Sumatera
Karang, 25114 Padang SumBar, Indonesia
Contact No.(s) : (Tel.) 31678, 55332
(Fax) 55475, 51591
Subject of Interest
Major : Behavior, ecophysiology, growth
Minor : Coral reefs, seaweeds, stock assessment

235. Dr. Methil N. **Kutty**
Consultant, PRASADAM, 5/383 Puthur
Palakkad 678001, Kerala State, India
Contact No.(s) : (Tel.) 28882

Subject of Interest
Major : Ecophysiology, freshwater aquaculture, training
Minor : Brackishwater aquaculture, carps, curriculum development

236. Mr **Kyaw** Myint **Oo**
Lecturer, Department of Zoology,
Yangon University
University PO, Yangon, Myanmar
Contact No.(s) : (Tel.) 62113
Subject of Interest
Major : Environmental studies
Minor : Pesticides, pollution

237. Mr Philippe E. **La Hausse De Lalouviere**
Project Manager
Management and Development Company Ltd.
Gentilly, Moka, Mauritius
Contact No.(s) : (Tel.) 4334225
(Fax) 4334145
Subject of Interest
Major : Freshwater aquaculture
Minor : Ecology

238. Mr. Maciu **Lagibalavu**
Principal Fisheries Officer, Fisheries
Department, Ministry of Primary Industries,
Box 358, Suva, Fiji
Contact No.(s) : (Tel.) 361122
(Fax) 351079
Subject of Interest
Major : Population genetics
Minor : Computer methods, policy

239. Dr. Kuldeep K. **Lal**
Scientist, National Bureau of Fish Genetic
Resource 2, Bagambari Housing Scheme,
Shivnagar, Allahapur
Allahabad 211006 U.P., India

Subject of Interest
Major : Reproductive physiology
Minor : Biotechnology, cryopreservation

240. Dr. T.J. **Lam**
Professor, Department of Zoology
National University of Singapore
Lower Kent Ridge, Singapore 0511
Contact No.(s) : (Tel.) 7722692
(Fax) 7792486
Subject of Interest
Major : Endocrinology, induced spawning, seed production
Minor : Broodstock, nutrition

241. Mr. Danilo B. **Largo**
Section Head,
Marine Biology Section
University of San Carlos
Talamban, Cebu City 6000,
Philippines
Contact No.(s) : (Tel.) 85021 to 4
Subject of Interest
Major : Growth, seaweeds, stock assessment
Minor : Conservation, coral reefs, microbiology phytoplankton

242. Mr. Jerome **Lazard**
Director, Programme aquaculture et peche
Centre de cooperation international en
recherche agronomique pour le development
c/o Cemagref, BP 5095
34033 Montpellier, Cedex 1, France
Contact No.(s) : (Tel.) 670046300
(Fax) 67635795
Subject of Interest
Major : Freshwater aquaculture, production systems, tilapias
Minor : Brackishwater aquaculture, cages, pond managment

243. Mr. Le **Tanh Duong**
Deputy Director
Mekong Delta Farming Systems Research
and Development Centre
University of Cantho, Cantho, Vietnam
Contact No.(s) : (Tel.) 24517
(Fax) 23407
Subject of Interest
Major : Freshwater aquaculture, integrated farming systems, rice-fish farming systems
Minor : Environmental studies, socioeconomics, software

244. Mr. Le **Thanh Hung**
Deputy Dean, The Faculty of Fisheries
The University of Agriculture and Forestry
Thu Duc, Ho Chi Minh City, Vietnam
Subject of Interest
Major : Brackishwater aquaculture, crustaceans, integrated farming systems
Minor : Nutrition

245. Dr. Le **Thanh Luu**
Director, Transfer of Technology Centre
Research Institute for Aquaculture
No. 1 Dinh Bang, Tien Son, Ha Bac, Vietnam
Contact No.(s) : (Tel.) 271368
(Fax) 59257 (C/o FAO Hanoi)
Subject of Interest
Major : Brackishwater aquaculture, crustaceans, feed technology, nutrition
Minor : Farming systems research, freshwater aquaculture, integrated farming systems

246. Mr. Le **Xuan Sinh**
Researcher, Faculty of Fisheries
University of Cantho, Cantho, Vietnam
Subject of Interest
Major : Brackishwater aquaculture, economics
Minor : Farming systems research

247. Mr. Marc **Legendre**
ORSTOM, 911 Avenue Agropolix, BP 5065
34032 Montpellier, Cedex 1, France
Contact No.(s) : (Tel.) 67046360
(Fax) 67547800
Subject of Interest
Major : Catfishes, genetics resources, reproductive physiology
Minor : Growth, production systems, tilapias

248. Dr. James **L. Lester**
Interim Dean/Professor
University of Houston-Clear Lake
2700 Bay Area Boulevard, Houston, Texas
77058, USA
Contact No.(s) : (Tel.) 2833700
(Fax) 2833707
Subject of Interest
Major : Population genetics, quantitative genetics
Minor : Crustaceans, genetic resources, tilapias

249. Mr. **Li Kangmin**
Associate Professor
Asian Pacific Regional Research & Training
Centre for Integrated Fish Farming, Wuxi
214081, China
Contact No.(s) : (Tel.) 601424, 601434
(Fax) 203304
Subject of Interest
Major : Environmental studies
Minor : Phytoplankton, rice-fish farming systems

250. Dr. Li **Sifa**
Professor, Department of Aquaculture
Shanghai Fisheries University
334 Jun Gong Road, Shanghai 200090, China
Contact No.(s) : (Tel.) 5431090
(Fax) 5434358

Subject of Interest
Major : Conservation, freshwater aquaculture, genetic resources

251\. Dr. Clive **Lightfoot**
Senior Scientist, ICLARM
MCPO Box 2631, 0718 Makati
Metro Manila, Philippines
Contact No.(s) : (Tel.) 8189283, 8180466
(Fax) 8163183
Subject of Interest
Major : Farming systems research, intergrated farming systems, rice-fish farming systems
Minor : Ecology, environmental impact, socioeconomics

252\. Dr. C. Kwei **Lin**
AIT, GPO Box 2754, Bankok 10501, Thailand
Contact No.(s) : (Tel.) 5245458
(Fax) 5246200
(Tlx.) 84276 TH
Subject of Interest
Major : Pond dynamics, tilapias, water quality
Minor : Brackishwater aquaculture, environmental studies

253\. Dr. Otom ar **Linhart**
Associate Professor
Reproduction et genetique des poissons
Laboratoire de biologie cellularine CHRU
49033 Angers Cedex 1, France
Contact No.(s) : (Tel.) 41338105, 41354735
(Fax) 41354138
Subject of Interest
Major : Carps, catfishes, freshwater aquaculture
Minor : Broodstock, cryopreservation, genetics, population genetics, reproductive physiology

254\. Dr. David C. **Little**
Assitstant Professor
Division of Agricultural & Food Engineering
AIT, GPO Box 2754, Bankok 10501, Thailand

Contact No.(s) : (Tel.) 5245473
(Fax) 5245473

Subject of Interest

Major : Farming systems research, integrated farming systems, tilapias

Minor : Induced spawning, rice-fish farming systems, waste-fed

255. Dr. David Timothy **J. Littleweed**
Research Fellow, Department of Palaeontology,
National History Museum, Cromwell Rd.
London SW7 5BD, UK
Contact No.(s) : (Tel.) 9389228
(Fax) 9389277

Subject of Interest

Major : Bivalves, marine acquaculture, taxonomy

Minor : Pathology

256. Ms. Hui **Liu**
Engineer, Beijing Aquatic Science Institute
Beijing 100075, China
Contact No.(s) : (Tel.) 7226669
(Fax) 3014181

Subject of Interest

Major : Freshwater aquaculture, genetics, reproductive physiology

Minor : Carps, cryopreservation, tilapias

257. Dr. Khalid P. **Lone**
Associate Professor, Department of Zoology
Punjab University, Quaid-I-Azam Campus
Lahore 54590, Pakistan
Contact No.(s) : (Tel.) 864028

Subject of Interest

Major : Biochemistry, bioenergetics, endocrinology, growth, nutrition, reproductive physiology

Minor : Brackishwater aquaculture, carps, environmental studies, marine aquaculture, pollution, tilapias

258. Mr. Nelson A. **Lopez**
Supervising Aquaculturist
Bureau of Fisheries & Aquatic Resources
860 Arcadia Bldg., Quezon Ave.
Quezon City, Philippines
Contact No.(s) : (Tel./Fax) 987805
Subject of Interest
Major : Policy, resource economics
Minor : Milkfish

259. Mr. Alessandro **Lovatelli**
Mariculture Expert, FAO/AGUILA II
c/o FAO-Representative, Cerro Maika 115
Lomas de Chapultepec 11000, Mexico
Contact No.(s) : (Tel.) 6817866, 6837022 L.102
(Fax) 5205755
Subject of Interest
Major : Bivalves, marine aquaculture, training
Minor : Environmental studies, hatchery technology, trade

260. Dr. John S. **Lucas**
Associate Professor, Zoology Department
James Cook University, Townsville
Queensland 4811, Australia
Contact No.(s) : (Tel.) 814412
(Fax) 251570
Subject of Interest
Major : Bivalves, giant clams, marine aquaculture

261. Ms. Susan C. **Lumanlan-Mayo**
Research Staff
Bureau of Fisheries and Aquatic Resources
860 Arcadia Bldg., Quezon Ave.
Quezon City, Philippines
Contact No.(s) : (Tel./Fax) 987871
Subject of Interest
Major : Diseases, immunology, parasities, pathology
Minor : Brackishwater aquaculture, broodstock, faming systems research

262. Mr. Tula M. **Luna**
Apartado 18-0202, Lima, Peru
Contact No.(s) : (Tel.) 352035 L. 290
Subject of Interest
Major : Freshwater aquaculture, integrated farming systems, nutrition
Minor : Extension, feed technology, tilapias

263. Mr. Henery B.O. **Lungayia**
Assistant Research Officer
Kenya Marine and Fisheries Research
Institute, PO Box 81615,
Mombasa,
Kenya
Subject of Interest
Major : Catfishes, freshwater aquaculture, tilapias
Minor : Broodstock, endocrinology, pond management

264. Dr. Julie M. **Macaranas**
Post-Doctoral Fellow
Centrè for Biological Population Management
Queensland University of Technology
2 George St., Brisbane, Queensland 4001,
Australia
Contact No.(s) : (Tel.) 8641340
(Fax) 8641640
Subject of Interest
Major : Genetics, quantitative genetics
Minor : Biochemistry, physiology

265. Dr. K. A. **Narasinmaham**
Retired Principal Scientist
& Head, CMFRI
Anirudh, 1-8-1-Opposite
Sriramnagar P.O.
Kakinada-533003
India.
Contact No.(s) : (Tel.) 0884-62817

Subject of Interest

Major : Marine fisheries resources and mollusc culture (oyster, mussel, clam, marine pearl, and abalone culture)

Minor : Biodiversity, benthos and environmental impact assessment of living resources.

Subject of Consultancy

Marine fisheries resources management
Aquaculture of Oyster. Clam, mussels, marine pearl and ablones
Biodiversity of coastal waters.
Benthos, plankton hydrography and nutrients.
Environmental impact assessment on living resources.

266. Dr. Mudanakudu C. **Nandeesha**
Asst. Professor
Department of Aquaculture, College of Fisheries, Mangalore 575002, Karnataka, India
Contact No.(s) : (Tel.) 27834
(Fax) 35069, 35079, 22395

Subject of Interest

Major : Hatchery technology, induced spawning, nutrition

Minor : Broodstock, farming systems research, freshwater aquaculture

267. Mr. Warwick J. **Nash**
Abalone Biologist, Sea Fisheries Division
Crayfish Point, Taroona, Tasmania 7053, Australia
Contact No.(s) : (Tel.) 277277
(Fax) 278035

Subject of Interest

Major : Marine aquaculture, stock assessment

268. Mr. A.K.V. **Nasser**
Scientist, Central Marine Fisheries Research Institute,
PB No. 2704,
Kochi 682031, India
Contact No.(s) : (Tel.) 369876
L.286

Subject of Interest
Major : Coral reefs, marine aquaculture, stock assessment
Minor : Computer methods, ecology, population genetics

269. Dr. Ted **Needham**
Director of Aquaculture Operations
British Columbia Packers Ltd.
PO Box 5000, Vancouver
British Columbia, Canada V 6B 4 AP
Contact No.(s) : (Tel.) 2413456
(Fax) 2771858
Subject of Interest
Major : Farming management, freshwater aquaculture, marine aquaculture
Minor : Cages, Diseases, fisheries enhancement

270. Mr. Michael B. **New**
Coordinator, ASEAN-EEC AADCP
PO Box 1006, Bangkok 10903, Thailand
Contact No.(s) : (Tel.) 5613022
(Fax) 5613023
Subject of Interest
Major : Crustaceans, development, nutrition
Minor : Brackishwater aquaculture, feed technology, freshwater aquaculture

271. Dr. Davies H. **Ng'ong'ola**
Senior Lecturer, Budna College of Agril.
PO Box 219, Lilongwe, Malawi
Contact No.(s) : (Tel.) 277222
(Fax) 277364
Subject of Interest
Major : Farming systems research, marketing, resource economics
Minor : Computer methods, economics, socioeconomics

272. Mr. David **Nguenga**
Research Officer, Institute of Animal Research
PO Box 255, Foumban, Cameroon
Contact No.(s) : (Tel.) 482124

Subject of Interest
Major : Integrated farming systems, seed production, water quality

273. Mr. **Nguyen** Anh Tuan
Dean, Faculty of Fisheries
University of Cantho, Cantho, Vietnam
Contact No.(s) : (Tel.) 20961, 20237
(Fax) 22101
Subject of Interest
Major : Freshwater aquaculture, pesticides, rice-fish farming systems

274. Mr. **Nguyen Duy** Can
Research Assistant
Mekong Delta Farming Systems Research
and Development Centre
University of Cantho, Cantho, Vietnam
Contact No.(s) : (Tel.) 24517
Subject of Interest
Major : Environmental studies, farming management, farming systems research
Minor : Integrated farming systems, modeling, rice-fish farming systems

275. Mr. **Nguyen** Quang Tuyen
Researcher, Agril. Systems Program
Faculty of Agril., Chiangmai University
Chiangmai 50002, Thailand
Contact No.(s) : (Fax) 210000
Subject of Interest
Major : Freshwater aquaculture, integrated farming systems, rice-fish farming systems
Minor : Environmental impact, farming systems research, resource economics, stock assessment, training

276. Mr. **Nguyen Van Sanh**
Mekong Delat Farming System Research
and Development Centre
University of Cantho, Cantho, Vietnam
Contact No.(s) : (Tel.) 24517

Subject of Interest
Major : Integrated farming systems, modeling socioeconomics
Minor : Marketing, production systems, rural sociology

277. Mr. Plutomeo M. **Nieves**
Asstt. Professor, Bicol University
Regional Institute of Fisheries
Technology
Tabaco, Albay 4916, Philippines
Subject of Interest
Major : Diseases, hatchery technology, nutrition

278. Dr. Chareen **Nitithamyong**
Lecturer, Department of Marine Science
Faculty of Science, Chulalongkorn University
Prathum-one, Bangkok 10330, Thailand
Contact No.(s) : (Tel.) 2516968, 1511951
(Fax) 2511951
Subject of Interest
Major : Bioenergetics, ecology, foodchains, stock assessment
Minor : Foodchains, software, statistics

279. Mr. Anthony A. **Nlewadim**
Research Officer
Nigerian Institute for Oceanography
and Marine Research
PMB 5122, Port Harcourt, Rivers State,
Nigeria
Subject of Interest
Major : Tilapias
Minor : Barckishwater aquaculture, integrated farming systems

280. Mr. Christopher G. **Nugent**
Chief Technical Advisor, FAO
PO Box 3730, Harare, Zimbabwe
Contact No.(s) : (Tel.) 703820
(Fax) 729563
Subject of Interest
Major : Extension, integrated farming system, reservoirs

281. Mr. Francis **Oduro-Boateng**
Project Officer,
Technology Counsultancy Centre
University of Science & Technology,
Kumasi, Ghana
Contact No.(s) : (Tlx.) 2555 UST GH
Subject of Interest
Major : Freshwater aquaculture, tilapias
Minor : Cages

282. Mr. Paul C. **Ofojekwu**
Lecturer / Research Fellow, Department of Zoology
University of Jos, PMB 2084, Jos Plateau
State, Nigeria
Subject of Interest
Major : Hatchery technology, nutrition, pond management

283. Mr. Patrick K. **Ofori-Danson**
Research Officer, Institute of Aquatic
Biology
PO Box 38, Achimota, Ghana
Contact No.(s) : (Tel.) 775511 (Accra)
(Fax) 772642 (Accra)
Subject of Interest
Major : Ecology, growth, stock assessment
Minor : Population genetics

284. Mr. Nyarko D. **Ohemeng**
Deputy Director, Irrigation Development
Authority
PO Box M 154, Accra, Ghana
Contact No.(s) : (Tel.) 662050
Subject of Interest
Major : Freshwater aquaculture, rice-fish farming systems
Minor : Environmental studies

285. Dr Daniel O. **Okeyo**
Senior Lectuter, Department of Zoology
Kenyatta University, PO Box 43844, Nairobi,
Kenya
Contact No.(s) : (Tel.) 810901 L: 305

Subject of Interest
Major : Freshwater aquaculture, integrated farming systems, tilapias
Minor : Conservation, growth, pond dynamics

286. Dr. Gideon C. **Okpokwasili**
Senior Lecturer, Department of Microbiology
University of Port Harcourt
PMB 5323, Port Harcourt, Nigeria
Subject of Interest
Major : Environmental impact, microbiology, water quality
Minor : Diseases, pollution, tilapias

287. Mr. Emmanuel A. **Okah**
Lecturer, Institute of Renewable Natural Resources, University of Sci. & Technology
Kumasi, Ghana
Contact No.(s) : (Tel.) 5353
(Fax) 2555
Subject of Interest
Major : Catfishes, induced spawning, nutrition
Minor : Finfish, freshwater aquaculture, seed production.

288. Dr. Eurico C. **Oliveira**
Professor, Institute of Biociencias
University of Sao Paulo
C. Postal 11461, S. Paolo 05499,
S.P., Brazil
Contact No.(s) : (Tel.) 8136944 L.2540
(Fax) 8154272
Subject of Interest
Major : Conservation, marine aquaculture, seaweeds
Minor : Ecology, ecophysiology, taxonomy

289. Mr. Victor A. **Omojola**
c/o Mr. E.O., Commerce Bank, PLC
PO Box 73715, Victoria Island, Lagos, Nigeria

Subject of Interest
Major : Carps, catfishes, freshwater aquaculture, intergrated farming systems, pond management, tilapias
Minor : Computer methods, ecology, hatchery technology, modeling, pond dynamics, software

290. Mr. Isaac T. **Omoniyi**
Lecturer II, Department of Zoology
Ondo State University, PMB 5363, Ado-Ekiti,
Nigeria
Subject of Interest
Major : Finfish, freshwater aquaculture, tilapias
Minor : Nutrition, seed production

291. Mr. Edosa **Omoregie**
Lecturer, Fisheries Research Unit
Department of Zoology
University of Jos, PMB 2084, Jos, Nigeria
Subject of Interest
Major : Nutrition, pollution, tilapias
Minor : Physiology

292. Mr. Christie A. **Onadeko**
Lecturer Grade I,
Department of Biological Sciences
University of Lagos, Akoka, Lagos, Nigeria
Contact No.(s) : (Tel.) 8211782
Subject of Interest
Major : Brackishwater aquaculture, feed technology, freshwater aquaculture, marine aquaculture
Minor : Farming systems research, lagoons, limnology, pond management

293. Mr. Boguslaw F. **Ostrowski**
Research Officer, Marine Sci. Department,
University Kebangsaan Malaysia, Kampus Sabah
Locked Bag 62, 88996 Kota Kinabalu, Sabah,
Malaysia

Subject of Interest
Major : Induced spawing, marine aqucaulture, reproductive physiology
Minor : Coral reefs, cryopreservation, ecology

294. Mr. Mark **Owusu-Frimpong**
Research Officer, Institute of Aquatic Biology
PO Box 38, Achimota, Ghana
Contact No.(s) : (Tel.) 775511 (Accra)
(Fax) 77655 (Accra)
Subject of Interest
Major : Catfishes, hatchery technology, tilapias
Minor : Behavior, extension, farming systems research

295. Mr. Benjy A. **Oyedeji**
Aquaculturist, PO Box 27268, Agodi Post Office
Ibadan, Oyo State, Nigeria
Contact No.(s) : (Tel.) 716016
Subject of Interest
Major : Freshwater aquaculture

296. Ms. Helida A. **Oyieke**
Senior Research Officer I
Kenya Marine & Fisheries Research Institute
PO Box 81615, Mombasa, Kenya
Contact No.(s) : (Tel.) 472245
(Fax) 472215
Subject of Interest
Major : Ecology, seaweeds, taxonomy
Minor : Marine aquaculture, site selection

297. Mr. Jose **O. Paclibare**
Aquaculturist II
Bureau of Fisheries & Aquatic Resources
860 Arcadia Bldg., Quezon Ave.
Quezon City, Philippines
Contact No.(s) : (Tel.) 987871
Subject of Interest
Major : Diseases
Minor : Microbiology

298. Prof. Velayutham **Padhanaban**
Prof. & Head
Department of Fish Processing & Technology
Fisheries College & Research Institute,
Tuticorin
Tamil Nadu-628008
India.
Contact No.(s) : (Tel.) 340554, 322354, 340554
(Fax) 340154
Subject of Interest
Major : Fish processing technology.
Minor : Value added fishery products.
Subject of Consultancy
Fish processing technology
Fish processing industry
Preparation of value added fishery products.

299. Prof. Abdullah Grande Soriano **Ponce**
King Abdulaziz University
Faculty of Marine Sciences
P.O. Box-1540, Obhur
Jeddah 21441,
Saudi Arabia.
Contact No.(s) : (Tel.) (02) 694-1427, (0063) 75-592-5128 (R)
Subject of Interest
Major : Aquaculture
Minor : General Fisheries.
Subject of Consultancy
Prawn and shrimp aquabusiness,
Tilapia and siganids culture,
Making project feasibility studies
Lay out and designs of aquaculture projects.

300. Mr. Acharya **Prodyoi**
29/1B, Christopher Road,
Calcutta-700046
India.
Subject of Interest
Major : Fish/Fishery biology
Minor : Inland, brackish, marine fishery and ecology

Subject of Consultancy
Aquaculture of inland fishes,
Farm management (Fisheries),
Socio-economic investigations in respect of manpower engaged in fish farming and allied indsutries, teaching, guiding dissertations, post graduate thesis, editing, developing syllabus.

301. Dr. P. **Pakshirajan**
Senior Technical Assistant
Department of Agril, and Cooperatives
Rm. 482A Krishi Bhavan
New Delhi 110001, India
Contact No.(s) : (Tel.) 388911 L.4682
(Fax) 386004
Subject of Interest
Major : Brackishwater aquaculture, farming management, integrated farming systems
Minor : Extension, farming systems research, marine aquaculture

302. Mr. K. **Palaniswamy**
Asstt. Development Officer
Technical Services Department
National Bank for Agril. & Rural Development
PB No. 6552, Worli, Bombay 400018, India
Contact No.(s) : (Tel.) 4924993
(Fax) 4949111
Subject of Interest
Major : Brackishwater aquaculture, freshwater aquaculture, integrated farming systems
Minor : Genetics, marine aquaculture, production systems

303. Ms. Rani **Palaniswamy**
Scientist, Central Marine Fisheries Research Institute
TRC of CMFRI, Tuticorin 628001, India
Contact No.(s) : 22274, 21515
Subject of Interest
Major : Broodstock, induced spawning
Minor : Bivalves

304. Mr. **Pan Feng**
Research Assistant
Freshwater Fisheries Research Centre
Baojieqiao, Wuxi, Jiangsu, China
Contact No.(s) : (Tel.) 601424
Subject of Interest
Major : Biotechnology, genetics, tilapias
Minor : Biochemistry, computer methods, hatchery technology

305. Ms. Ma. Josefa R. **Pante**
Univ. Researcher, Marine Sci. Institute
University of the Philippines
Contact No.(s) : (Tel.) 995071 to 9, 976061 to 9
(Fax) 974355
Subject of Interest
Major : Population genetics, quantitative genetics
Minor : Crustaceans, tilapias

306. Ms. Marcela S. **Pascual**
Researcher, Instituto de Biologia
Marine Y Pesquera
C.C. 104 (8520) San Antonio Oeste,
Argentina
Contact No.(s) : (Tel./Fax) 21002
Subject of Interest
Major : Bivalves
Minor : Ecology, marine aquaculture

307. Dr. Swapan K. **Paul**
Senior Scientist Officer
Fisheries Research Institute-Third
Fisheries Project
Santahar, Bogra, Bangladesh
Subject of Interest
Major : Broodstock, nutrition, reproductive physiology
Minor : Limnology, pollution, stock assessment

308. Dr. Daniel M. **Pauly**
Director, ICLARM
MCPO Box 2631, 0718 Makati
Metro Manila, Philippines
Contact No.(s) : (Tel.) 8189283, 8180466, 8175255
(Fax) 8163183
Subject of Interest
Major : Growth
Minor : Modeling

309. Mr. Malcolm G. **Pearce**
Technician
Queensland Department of Primary Industries
Northern Fisheries Research Centre, Box 5396
CMC
Cairns, Queensland, Austrilia
Subject of Interest
Major : Behavior, marine aquaculture, pond dynamics
Minor : Coral reefs, physiology, water quality

310. Dr. David J. **Penman**
Research Fellow, Institute of Aquaculture
University of Stirling, Stirling, FK9 4LA, UK
Contact No.(s) : (Tel.) 467901
(Fax) 472133
Subject of Interest
Major : Biotechnology, genetics
Minor : Freshwater aquaculture, genetics resources, population genetics

311. Dr. Julio E. **Perez**
Prof., Instituto Oceanografico de Venezuela
Universidad de Oriente
Apartado Postal 245, Cumanan, Venezuela
Contact No.(s) : (Tel.) 653606
(Fax) 653606
Subject of Interest
Major : Bivalves, genetics, tilapias
Minor : Genetic resources, population genetics, quantitative genetics

312. Dr. Peter C. **Phillips**
Research Scientist, Clark Atlanta University
Box 271, Atlanta, GA 30314, USA
Contact No.(s) : (Tel.) 8808811
(Fax) 8808522
Subject of Interest
Major : Pesticides, tilapias, wastewater (sewage) reuse
Minor : Detritus, integrated farming systems, waste-fed

313. Dr. Raul H. **Piedrahita**
Associate Professor
Department of Agril. Engineering
University of California, Davis, CA
956616, USA
Contact No.(s) : (Tel.) 7522780
(Fax) 7522640
Subject of Interest
Major : Engineering, modeling, water quality
Minor : Integrated farming systems, phytoplankton, pond dynamics

314. Dr. Robert S. **Pomeroy**
Research Scientist, ICLARM
MCPO Box 2631, 0718 Makati
Metro Manila, Philippines
Contact No.(s) : (Tel.) 818983, 8180466, 8175255
(Fax) 8163183
Subject of Interest
Major : Economics
Minor : Integrated farming systems, policy

315. Mr. Victor **Pouomogne**
Researcher, Fishculture Research Station
PO Box 255, Foumban, Cameroon
Subject of Interest
Major : Nutrition, tilapias

316. Mr. Anders F. **Povlsen**
Fisheries Biologist, FAO Field Office
PO Box 305, Madang, Papua New Guinea

Contact No.(s) : (Tel.) 823011
(Fax) 823306

Subject of Interest

Major : Ecology, freshwater aquaculture, reproductive physiology

Minor : Conservation, integrated farming systems, pollution

317. Dr. Howard W. **Powles**
Fisheries Specialist, CIDA-RNP
200 Portage, Hull, Quebec, Canada K 1A OG4
Contact No.(s) : (Tel.) 9976772
(Fax) 9534676

Subject of Interest

Major : Farming management, farming systems research, production systems

318. Dr. Mark **Perien**
Assoc. Scientist, ICLARM
MCOP Box 2631, 0718 Makati
Metro Manila, Philippines
Contact No.(s) : (Tel.) 8189283, 8180466, 8175255
(Fax) 8163183

Subject of Interest

Major : Computer methods, farming systems research, integrated farming systems

Minor : Modeling, pond dynamics, wastewater (sewage) reuse

319. Dr. Roger S.V. **Pullin**
Director, ICLARM
MCOP Box 2631, 0718 Makati
Metro Manila, Philippines
Contact No.(s) : (Tel.) 8189283, 8180466, 8175255
(Fax) 816183

Subject of Interest

Major : Freshwater aquaculture, genetics, integrated farming systems

Minor : Conservation, ecology, waste (sewage) reuse

320. Ms. Parveen **Quarrar**
Ph. D. Student/Research Assitant
Department of Biology, Univ. of New Brunswick
Bag Service 4511, Fredericton, N.B., Canada
E3B 6E 1
Contact No.(s) : (Tel.) 4534583 to 4
(Fax) 4533583
Subject of Interest
Major : Freshwater aquaculture, marine aquaculture, parasites
Minor : Ecology, farming management, immunology

321. Mr. Md. Khalilur **Rahman**
Scientific Officer, Fisheries Research
Institute, Riverine Station, Chandpur 3502
Bangladesh
Contact No.(s) : (Tel.) 3482, 3407
Subject of Interest
Major : Catfishes, induced spawning, reproductive physiology
Minor : Feed technology, freshwater aquaculture,

322. Dr. Krishen **Rana**
Research Lecturer, Institute of Aquaculture
Univ. of Stirling, Stirling FK9 4LA, UK
Contact No.(s) : (Tel.) 467901
(Fax) 472133
(Tlx.) 777759
Subject of Interest
Major : Biotechnology, cryopreservation, genetic resources
Minor : Hatchery technology, reproductive physiology, seed production

323. Dr. Carlos D. **Rangel**
Professor/Researcher
Universidad Autonoma de Baja California Sur
PO Box 18-B, La Paz 23081
Baja California Sur, Mexico
Contact No.(s) : (Tel.) 10755
(Fax) 11880

Subject of Interest
Major : Marine aquaculture
Minor : Bivalves

324. Mr. Krishna D.S. **Rao**
Senior Scientist, Reservoir Fisheries
Division, 1031, 80 Feet Rd., 4th Block,
Rajajinagar
Bangalore 560010, India
Contact No.(s) : (Tel.) 357213
Subject of Interest
Major : Limnology, reservoirs
Minor : Cages, phytoplankton, zooplankton

325. T. V. Gayathri **Rao**
Information Specialist, B.P. Nutrition
PO Box 5, Songkhla 90000, Thailand
Contact No.(s) : (Tel.) 321056, 321057
(Fax) 321058, 321055
Subject of Interest
Major : Business development information, marine aquaculture
Minor : Broodstock, nutrition, pond management

326. Mr. Pascual M. **Remetio**
Project Coordinator, Lutheran World Services
PO Box 1133, Maputo, Mozambique
Contact No.(s) : (Tel.) 20261
(Fax) 20387
Subject of Interest
Major : Rice-fish farming systems
Minor : Farming systems research

327. Mr. Dirk J. R. **Reyntjens**
Aquaculturist, GOPA Consultants
Hindenburgring 18
Subject of Interest
Major : Farming systems research
Minor : Development, socioeconomics

328. Mr. Michael A. **Rimmer**
Senior Fisheries Biologist, Northern
Fisheries Centre
Box 5396, Cairns, Queensland, Australia
Contact No.(s) : (Tel.) 351560
(Fax) 351401
Subject of Interest
Major : Brackishwater aquaculture, ecology, marine aquaculture
Minor : Finfish, pond dynamics, zooplankton

329. Dr. Olukayode O. **Roberts**
National Coordinator
Nigerian Institute for Oceangraphy
and Marine Research
PMB 12729 Victoria Island, Lagos, Nigeria
Contact No.(s) : (Tel.) 630402
Subject of Interest
Major : Freshwater aquaculture, integrated farming systems, rural sociology
Minor : Cages, diseases, extension

330. Dr. Frederik C. **Roest**
Senior Fisheries Adviser
International Agril. Centre
PO Box 88, 6700 AB Wageningen, The Netherlands
Contact No.(s) : (Tel.) 90247
(Fax) 18852
Subject of Interest
Major : Development, environmental studies, marine aquaculture

331. Mr. Lanchipa G.F. **Rojas**
Tecnico Analista, Instituto de Fomento Pesquero
PO Box 170, Desiderio Garcia S/N
Barrio Industrial Sitio, No. 2, Iquique, Chile
Contact No.(s) : (Tel.) 413579, 422216
(Fax) 411944

Subject of Interest
Major : Brackishwater aquaculture, farming systems research, freshwater aquaculture, hatchery technology, integrated farming systems, marine aquaculture, modeling, production systems, seaweeds, seed production, site selection, water quality
Minor : Biochemistry, development, ecology, engineering, environmental impact, environmental studies, feed technology, habitat restoration, information, pathology

332. Ms. MaRowena R. **Romana-Eguia**
Research Associate
SEAFDEC Aquaculture Department
Binangonan Freshwater Station
Binangonan, Rizal 1903, Philippines
Contact No.(s) : (Tel.) 22125, 9230201 to 3 (Manila)
(Fax) 81340 (Iloilo)
Subject of Interest
Major : Freshwater aquaculture, genetics, tilapias
Minor : Population genetics, quantitative genetics

333. Mr. Ivar **Ronnestad**
Scientist, Zoological Laboratory
Allegt. 41 N-5007, Bergen, Norway
Contact No.(s) : (Tel.) 212246
Subject of Interest
Major : Bioenergetics, marine aquaculture, physiology
Minor : Biochemistry, modeling

334. Mr. Bob **Rosenberry**
Editor/Publisher, Aquaculture Digest
9434 Kearny Mesa Rd., San Diego, CA 92126
USA
Contact No.(s) : (Tel.) 2716354
(Fax) 2710324
Subject of Interest
Major : Information, shrimps

335. Dr. Lindsay G. **Ross**
Lecturer, Institute of Aquaculture
University of Stirling, Stirling FK9
4LA, UK
Contact No.(s) : (Tel.) 467882
(Fax) 471233
Subject of Interest
Major : Bioenergetics, environmental studies, physiology
Minor : Computer methods, curriculum development, site selection

336. Dr. Shmuel **Rothbard**
Chief Biologist, YAFIT Laboratory
Fish Breeding Centre, Gan Shmuel 38810, Israel
Contact No.(s) : (Tel.) 320008
(Fax) 320404
Subject of Interest
Major : Broodstock, genetics, induced spawning

337. Dr. Kenneth R. **Ruddle**
Prof. Matsugaoko-cho 11-20
Nishinomiya-shi, Hyogo-ken 662, Japan
Contact No.(s) : (Tel./Fax) 2904
Subject of Interest
Major : Ecology, farming systems research, integrated farming systems
Minor : Economics, modeling

338. Dr. Christopher Paul **Matheus**
Northfield House
Cheselbourne
Nr. Dorchester
Dorset, OTZ INT
England
Contact No.(s) : (Tel+.) 44-1258837133
Subject of Interest
Major : Fisheries management assessment, Economics.
Minor : Aquaculture

Subject of Consultancy
Fisheries management, assessment and aquaculture management in coastal zones (Any where in the Indian subcontinent South or North East Asia)

339. Dr. Rabinarayan **Mishra**
Assistant Professor,
Fishery Engineering, College of Fisheries,
Rangailunda, Berhampur (Ganjam)
Orissa-760007, India
Contact No.(s) : (Tel.) 0680-282 456, 0674-430742 (R)
Subject of Interest
Major : Fishery Engineering (Gear engg, Craft engg., Fishing tech. Aquaculture engg, Refrigeration engg.)
Minor : Fish Processing technology, Fishery microbiology.
Subject of Consultancy
Gear Engineering, Fishing Technology, Aquaculture engineering, Fish processing technology.

340. Dr. Marcel A.M. **Machiels**
Lecturer, Department of Fish Culture
and Fisheries Wageningen Agricultural
University, PO Box 338
6700 AH Wageingen, The Netherlands
Contact No.(s) : (Tel.) 83801
Subject of Interest
Major : Reservoirs, statistics, stock assessment
Minor : Computer methods, foodchains, modeling

341. Dr. Kevan L. **Main**
Director
Central for Tropical & Subtropical
Aquaculture, The Oceanic Institute,
Makapuu Pt.
Waimanalo, HI 96795, USA
Contact No.(s) : (Tel.) 2597951
(Fax) 2598395
Subject of Interest
Major : Brackishwater aquaculture, freshwater aquaculture, mairne aquaculture
Minor : Diseases, genetics, giant clams

342. Dr. Graham C. **Mair**
Research Consultant, Freshwater Aquaculture
Centre, Central Luzon State University,
Munoz, Nueva Ecija 3120, Philippines
Contact No.(s) : (Tel./Fax) 3017323
Subject of Interest
Major : Development, genetics, tilapias
Minor : Biotechnology, induced spawning, population genetics

343. Mr. Alfred **O. Maluwa**
Fisheries Research Officer
Central & Northern Regions Fish
Farming Project
PO Box 700, Mzuzu, Malawi
Contact No.(s) : (Tel.) 332794, 332692
(Fax) 335051
Subject of Interest
Major : Induced spawning, integrated farming systems, tilapias
Minor : Freshwater aquaculture, hatchery technology, seed production

344. Mr. Jose V. **Mazano**
Professor, College of Fisheries,
Bicol University
Tabaco, Albay 4511, Philippines
Contact No.(s) : (Tel.) 672
Subject of Interest
Major : Brackishwater aquaculture

345. Mr. Rafael R. **Mapunda**
Principal, Freshwater Fisheries Institute
Box 1213, Mwanza, Tanzania
Contact No.(s) : (Tel.) 2894
Subject of Interest
Major : Biotechnology, farming systems research
Minor : Hatchery technology, tilapias, training

346. Mr. Jimmy T. **Masagca**
Assistant Professor,
College of Agriculture and Fisheries
Virac, Catanduanes 4800, Philippines
Contact No.(s) : (Tel.) 473
Subject of Interest
Major : Environmental Studies, genetics
Minor : Genetic resources

347. Dr. Peter B. **Mather**
Lecturer, School of Life Science
Queęnsland University of Technology
2 George St., Brisbane 4001, Australia
Contact No.(s) : (Tel.) 8641737
(Fax) 8641640
Subject of Interest
Major : Freshwater aquaculture, genetics, tilapias
Minor : Brackishwater aquaculture, genetic resources, stock assessment

348. Dr. Christopher P. **Mathews**
International Consultant, Northfield House
Cheselbourne Near Dorchester, Dorset DT2
7NT, UK
Contact No.(s) : (Tel.) 77133
Subject of Interest
Major : Policy, socioeconomics, stock assessment

349. Dr. Toshihiko **Matsusato**
Prof. Officer, FAO-FIRI
Rm. F-236, Via delle Terme di Caracalla
00100 Rome, Italy
Contact No.(s) : (Tel.) 57976473
Subject of Interest
Major : Diseases, parasites, pathology
Minor : Immunology, microbiology, pesticides

350. Dr. Brendan J. **McAndrew**
Senior Lecturer, Institute of Aquaculture
University of Stiriling
Stiriling FK9 4LA, Scotland, UK

Contact No.(s) : (Tel.) 467884
(Fax) 473211

Subject of Interest

Major : Genetic resources, genetics, tilapias

Minor : Cryopreservation, population genetics taxonomy

351. Mr. Carlos E. **Medina-Reyna**
Research Asst, Apartado Postal 113
70670 Salina Cruz, Oaxaca, Mexico

Subject of Interest

Major : Brackishwater aquaculture, environmental studies, farming management, statistics, stock assessment, water quality, zooplankton

Minor : Computer methods, feed technology, hatchery technology, induced spawning, nutrition reproductive physiology, seed production, tilapias

352. Mr. Joaquin P. **Mellado**
Lecturer / Researcher, Apartado Postal 742
Gyaymas, Sonora, Maxico

Subject of Interest

Major : Finfish, marine aquaculture, pathology

353. Mr. Khalid A. **Memon**
Managing Director
Aquaculture Technologies (Pvt.) Ltd.
18 B, Coopretative Housing Society, Makli
73150 Thatta, Sindh, Pakistan

Contact No.(s) : (Tel.) 267
(Fax) 839

Subject of Interest

Major : Carps, freshwater aquaculture, freshwater prawns, hatchery technology

Minor : Farming management, feed technology, tilapias

354. Mr. Kriangsak **Meng-Umpan**
Head of Department / Instructor
Department of Fisheries Technology
Faculty of Agricultural Production
Maejo Institute of Agriculture Technology
Sansai, Chiangmai 50290, Thailand

Contact No.(s) : (Tel.) 498178
(Fax) 498862

Subject of Interest
Major : Endocrinology, induced spawning
Minor : Integrated farming systems

355. Ms. Zuridah **Merican**
Post-Graduate Student, Department of Zoology,
National University of Singapore
Lower Kent Ridge, Singapore 0511
Contact No.(s) : (Tel.) 7722708
(Fax) 7792486

Subject of Interest
Major : Nutrition
Minor : Crustaceans, Finfish

356. Dr. Bernard O. **Mgbenka**
Senior Lecturer, Department of Zoology
University of Nigeria, Naukka, Enugu State,
Nigeria

Subject of Interest
Major : Catfishes, freshwater aquaculture, nutrition
Minor : Diseases, environmental impact, feed technology, hatchery technology, pollution, wastefed, water quality

357. Mr. Wilson **Mhlanga**
Ecologist, Lake Kariba Fisheries Research
Institute PO Box 75, Kariba, Zimbabwe
Contact No.(s) : (Tel.) 2936 to 7
(Fax) 2938

Subject of Interest
Major : Cages, freshwater aquaculture
Minor : Pollution

358. Dr. Jean-Claude **Micha**
Prof. Unite d'ecologie des eaux douces
Facultes Universitaires
rue de Bruxelles 61, B-5000 Namur, Belgium
Contact No.(s) : (Tel.) 724361
(Fax) 724420

Subject of Interest
Major : Freshwater aquaculture, integrated farming systems, stock assessment
Minor : Environmental studies, nutrition, pollution

359. Dr. Ana **Milstein**
Researcher, Fish & Aquaculture Research
Station, Dor M.P. Hof Hacarmel 30820,
Israel
Contact No.(s) : (Tel.) 390652
(Fax) 993998
Subject of Interest
Major : Freshwater aquaculture, limnology, water quality
Minor : Foodchains, pond dynamics, statistics

360. Mr. **Min Kuanhong**
Trainer
Asia Pacific Regional Research & Training
Centre for Integrated Fish Farming, Wuxi
214081, China
Contact No.(s) : (Tel.) 601424
(Fax) 603304
Subject of Interest
Major : Integrated farming systems, extension, training
Minor : Curriculum development, farming systems research, freshwater aquaculture

361. Ms. S. Suzanne M. **Mingoa**
University Researcher, Marine Science
Institute, University of the Philippines
Diliman 1101
Quezon City, Philippines
Contact No.(s) : (Tel.) 98967 to 9 L. 7417
(Fax) 974355
Subject of Interest
Major : Ecophysiology, giant clams, marine aquaculture
Minor : Broodstock, extension, growth

362. Mr. Dan **Mires**
Chief of Aquaculture Division,
Fisheries Department, Ministry of Agriculture
Hakyria, Tel Aviv 61070, Israel
Contact No.(s) : (Tel.) 6971546
(Fax) 6971451
Subject of Interest
Major : Freshwater aquaculture, statistics
Minor : Tilapias

363. Dr. Samir Y. **Mishrigi**
Asst. Prof./Head of Aquaculture Department
PO Box 1489, Khartoum, Sudan
Subject of Interest
Major : Freshwater aquaculture, pond management, tilapias
Minor : Bioenergetics, ecology, ecophysiology

364. Dr. Arunabha **Mitra**
Associate Professor
Agricutural Engineering Department
Indian Institute of Technology
Kharagpur 721302, India
Contact No.(s) : (Tel.) 2221 to 4
(Fax) 2303
Subject of Interest
Major : Brackishwater aquaculture, cryopreservation, wastewater (sewage) reuse
Minor : Environmental studies, hatchery technology, integrated farming systems

365. Mr. Andrew **Mittelman**
Agro and Community Forestry Project,
Save the Children
PO Box 49, Nakhon, Sawan 60000, Thailand
Contact No.(s) : (Tel.) 221385
(Fax) 226213
Subject of Interest
Major : Habitat restoration, integrated farming systems, policy
Minor : Behavior, development, ecology

366. Dr. M.V. **Mohan**
Assistant Prof. Rice Research Station
Kerala Agricultural University, Vyttila
Kochi 682019, India
Contact No.(s) : (Tel.) 45063
Subject of Interest
Major : Farming systems research, pond dynamics, tilapias
Minor : Bioenergetics, brackishwater aquaculture, stock assessment

367. Dr. Krit **Mongkonpunya**
Assoc. Prof. Department of Zoology
Faculty of Science, Kasetsart University
Bangkok 10900, Thailand
Contact No.(s) : (Tel.) 5791022
(Fax) 5790514
Subject of Interest
Major : Conservation, reproductive physiology
Minor : Biotechnology, catfishes

368. Mr. Danilo **C. Monje**
Instructor, Freshwater Aquaculture Center
Central Luzon State University
Munoz, Nueva Ecija 3120, Philippines
Contact No.(s) : (Tel.) 3017323
Subject of Interest
Major : Hatchery technology, induced spawning, integrated farming systems
Minor : Cages, carps, tilapias

369. Ms. Priscila M. **Monje**
Instructor / Research Biologist
Freshwater Aquaculture Center
Central Luzon State University
Munoz, Nueva Ecija 3120, Philippines
Contact No.(s) : (Tel.) 3017323
Subject of Interest
Major : Freshwater aquaculture
Minor : Integrated farming systems, pesticides, post-harvest technology, water quality

370. Mr. Carlos Manuel B. **Montenegro**
Calle 18 A N. 41-82, Pasto Narino, Colombia
Contact No.(s) : (Tel.) 223539, 221933
Subject of Interest
Major : Pond management
Minor : Freshwater aquaculture

371. Dr. Jacques J. **Moreau**
University Prof.
Fisheries Department
ENSA, 145 Avenue de Muret, 31076 Toulouse,
France
Contact No.(s) : (Tel.) 428398
(Fax) 423029
Subject of Interest
Major : Ecology, environmental impact, stock assessment
Minor : Foodchains, growth, limnology, statistics, tilapias

372. Mr. Pierre **Morissens**
Programme Chief,
Institut des Savanes
BP 621, Bouake 01, Cote d'Ivoire
Contact No.(s) : (Tel.) 632170
(Fax) 632045
Subject of Interest
Major : Brackishwater aquaculture, production systems, -tilapias
Minor : Farming systems research, foodchains, pond dynamics

373. Mr. Julio **Moscoso**
Los Pinos 259,
Urb. Camacho, La Molina,
Peru
Contact No.(s) : (Tel.) 370177
(Fax) 378289
Subject of Interest
Major : Tilapias, wastewater (sewage) reuse
Minor : Farming systems research, integrated farming systems

374. Mr. Shaun M. **Moss**
Graduate Student, Department of Zoology
University of Hawaii, Honolulu, HI 96822,
USA
Contact No.(s) : (Tel.) 9568617
(Fax) 9569812
Subject of Interest
Major : Brackishwater aquaculture, ecology
Minor : Pond dynamics

375. Mr. Orton V. **Msiska**
Principal Scientific Officer
Department of Research & Environmental
Affairs, PO Box 30745, Lilongwe, Malawi
Contact No.(s) : 734011
Subject of Interest
Major : Catfishes, quantitative genetics, tilapias
Minor : Conservation, farming systems research, genetic resources

376. Dr. John L. **Munro**
Director, ICLARM
MCOP Box 2631, 0718 Makati
Metro Manila, Philippines
Contact No.(s) : (Tel.) 8189283, 8180466, 8175255
(Fax) 8163183
Subject of Interest
Major : Coral reefs, giant clams, production systems
Minor : Computer methods, ecology, marine aquaculture, stock assessment

377. Mr. Masayoshi **Murakoshi**
Senior Researcher
Okinawa Preectural Sea Farming Center
853-1, Ohama, Motobu Town, Okinawa 905-02,
Japan
Contact No.(s) : (Tel.) 475411
(Fax) 475412
Subject of Interest
Major : Bivalves, reproductive physiology, seed production
Minor : Coral reefs, marine aquaculture

378. Dr. Shivananda H. **Murthy**
Assistant Prof. Department of Aquaculture
College of Fisheries, PB No. 527
Mangalore 575002, Karnataka, India
Subject of Interest
Major : Freshwater aquaculture, nutrition
Minor : Brackishwater aquaculture

379. Dr. Natarajan **Murugan**
Prof., Department of Zoology
Madurai College (Autonomous)
Madurai 625011, India
Contact No.(s) : (Tel.) 22454
Subject of Interest
Major : Freshwater aquaculture
Minor : Zooplankton

380. Mr. Ahmed **Mustafa**
Ph.D. Student/Research Assistant
Department of Biology, University of New
Brunswick Bag Service 45111, Frederiction,
N.B., Canada E3B 6E1
Contact No.(s) : (Tel.) 4534583 to 4
(Fax) 4533583
Subject of Interest
Major : Freshwater aquaculture, marine aquaculture, parasites
Minor : Ecology, farming management, training

381. Mr. Jaston T. **Mutambo** (Deceased)
Aquaculture Research Officer
National Aquaculture Centre
PO Box 44, Domasi, Malawi
Contact No.(s) : (Tel.) 531215
(Fax) 523058 (Attn.: MAGFAD)
Subject of Interest
Major : Pond management
Minor : Freshwater aquaculture

382. Mr. P. **Muthiah**
Scientist, Central Marine Fisheries
Research Institute 90 North Beach Rd.,
Tuticorin-1, India
Subject of Interest
Major : Bivalves, marine aquaculture
Minor : Induced spawning

383. Ms. Uthairat **Na-Nakorn**
Assoc. Prof. Department of Aquaculture
Faculty of Fisheries, Kasetsart University
Bangkok 10903, Thailand
Contact No.(s) : (Tel.) 5792924
(Fax) 5795579
Subject of Interest
Major : Biotechnology, genetics, quantitative genetics
Minor : Induced spawning

384. Dr. P. **Nammalwar**
Senior Scientist/Project Leader,
Regional Centre of Central Marine Fisheries Research
Institute, Marine Fisheries Post, Mandapam
Ramanathapuram District
Tamil Nadu 623520, India
Contact No.(s) : (Tel.) 39443, 39456
Subject of Interest
Major : Finfish, marine aquaculture, seed production
Minor : Broodstock, integrated farming systems, pollution

385. Dr. **Nammalwar** Pakshi Ratan
Sr. Scientist
Central Marine Fisheries Research Institute (ICAR)
68/1, Greams Road, Madras-600006, India
Contact No.(s) : (Tel.) 044-8253299, 419414 (R)
(Fax) 044-8254252
Subject of Interest
Major : Marine fisheries, marine aquaculture, research and development, marine pollution and protection
Minor : Endangered marine mammals protection, marine biodiversity, conservation and management.

Subject of Consultancy
Survey assessment and monitoring the resources characteristics of pelagic and demersal marine finfishes, and estimation of potential and maximum sustainable yield.
Marine finfihes (milk fish grey mullets) culture, production and farm management and seed production technology.
Marine pollution monitoring protection and environmental management.
Biodiversity conservations.

386. Mr. Jagath W. **Rupasinghe**
Lecturer, Department of Fisheries Biology
Faculty of Science, University of Ruhuna
Matara, Sri Lanka
Contact No.(s) : (Tel.) 2681
Subject of Interest
Major : Economics, marketing, post-harvest technology
Minor : Resource economics, socioeconomics

387. Dr. Suleiman O.E. **Sadiku**
Ph.D. Student, Institute of Aquaculture
University of Stirling, Stirling FK9 4LA, UK
Contact No.(s) : (Fax) 472133
Subject of Interest
Major : Freshwater aquaculture, integrated farming systems, nutrition
Minor : Computer methods, statistics, stock assessment

388. Mr. Shyamalendu B. **Saha**
Senior Scientific Officer
Riverine Sub-station, Fisheries Research Institute
PO Box 08, Rangamati 4500, Bangladesh
Contact No.(s) : (Tel.) 2159
Subject of Interest
Major : Freshwater aquaculture, limnology, pond dynamics
Minor : Stock assessment

389. Mr. P. **Sahadevan**
Officer, The Waterbase Ltd., W-16/1681
Ramamurthy Nagar, Nellore, Andra Pradesh
India

Contact No.(s) : (Tel.) 28461,
23633
(Fax) 23955
Subject of Interest
Major : Brackishwater aquaculture, nutrition, pond management
Minor : Crustaceans, seed production

390. Mr. Wichien **Sakaras**
Trang Coastal Aquacultures Station,
Pakmeng Beach
Sikao District, Trang 92150, Thailand
Contact No.(s) : (Tel.) 212177
Subject of Interest
Major : Brackishwater aquaculture
Minor : Cages, finfish, nutrition

391. Mr. T.M. **Sankaran**
Assistant Professor
Head of Department of Management Studies
College of Fisheries,
Kerala Agril University
Panagad, Cochin 682506, India
Contact No.(s) : (Tel.) 805274
Subject of Interest
Major : Statistics, stock assessment
Minor : Modeling

392. Mr. Alejandro E. **Santiago**
Station Manager, SEAFDEC Aquaculture Department
Binangonan Freshwater Station
Binangonan, Rizal 1903, Phillipines
Contact No.(s) : (Tel.) 22125,
9230201 to 04 (Manila)
(Fax) 81340 (Iloilo)
Subject of Interest
Major : Ecology, limnology, phytoplankton
Minor : Environmental impact, wastewater (sewage) reuse, water quality

393. Dr. Amita **Saxena**
Asstt. Professor, College of Fisheries
Department of Aquaculture
G.B. Pantnagar University of Agriculture
and Technology
Pantnagar 263145, India
Subject of Interest
Major : Biochemistry, biotechnology, genetics
Minor : Nutrition, pathology

394. Dr. Erkki J. **Schwanck**
Research Officer, Department of Zoology
University of Stockholm, 106 91 Stockholm,
Sweden
Subject of Interest
Major : Behavior, reproductive physiology, tilapias

395. Dr. Lousie F. **Scura**
Resources Economist, C/o FINAQUA
Diagonal 6, 10-31, Zona 10, Guatemala
Contact No.(s) : (Tel.) 319683
(Fax) 346573
Subject of Interest
Major : Environmental impact, resource economics, socioeconomics
Minor : Economics, marketing, training

396. Mr. Ruben C. **Sevilleja**
Director/Assoc. Professor
Freshwater Aquaculture Centre
Central Luzon State University
Muñoz, Nueva Ecija 3120, Philippines
Contact No.(s) : (Tel./Fax) 3017323
Subject of Interest
Major : Economics, freshwater aquaculture, integrated farming systems
Minor : Farming systems research, rice-fish farming systems, socioeconomics

397. Dr. Meeransa S. **Shafee**
Professor, Fisheries Section
Institut agronomique et veterinaire
Hassan II
BP 6202 Rabat-Instituts, Rabat, Morroco
Contact No.(s) : (Tel.) 777439
(Fax) 778110
Subject of Interest
Major : Bivalves, lagoons, marine aquaculture
Minor : Bioenergetics, ecology, stock assessment

398. Mr. Maher M. **Shafik**
Senior Biologist
Institute of Inland Water and Fish Culture
10 Hassan Sabry St., Zamalek, Cairo, Egypt
Contact No.(s) : (Tel.) 3417298, 3401606
Subject of Interest
Major : Reproductive biology

399. Dr. Mohamed **Shariff**
Associate Professor
Faculty of Fisheries & Marine Science
Universiti Pertanian Malaysia, 43400 UPM Serdang
Selangor Darul Ebsan, Malaysia
Contact No.(s) : (Tel.) 9488246
(Fax) 9488246
Subject of Interest
Major : Diseases, immunology, parasities
Minor : Environmental studies, freshwater aquaculture, pathology

400. Dr. Colin C. **Shelley**
Director, Aquaculture
Department of Primary Industry & Fisheries
GPO Box 990, Darwin, NT 0801, Australia
Contact No.(s) : (Tel.) 894321
(Fax) 811475
Subject of Interest
Major : Farming systems research, finfish, induced spawning
Minor : Bivalves, policy, site selection

401. Dr. William L. **Shelton**
Prof., Zoology Department
University of Oklahoma, Norman, OK 73019, USA
Contact No.(s) : (Tel.) 3251058
Subject of Interest
Major : Carps, cryopreservation freshwater aquaculture genetics, growth, induced spawning, reproductive physiology, reservoirs, tilapias
Minor : Endorinology, water quality

402. Dr. Kerry L. **Shephard**
Senior Lecturer, Department of Life Sciences
The University of Buckingham
Buckingham, MK18 1EG, UK
Contact No.(s) : (Tel.) 814080
Subject of Interest
Major : Bioenergetics, ecophysiology, physiology
Minor : Environmental studies, pollution, water quality

403. Dr. Abdul Q. **Siddiqui**
Research Scientist, Fish Culture Project
Institute of Natural Resources and Environment
Kacst, PO Box 6088, Riyadh 11442, Saudi Arabia
Contact No.(s) : (Tel.) 4883555, 4883444 L. 603
(Fax) 4813611
Subject of Interest
Major : Freshwater aquaculture
Minor : Production systems, tilapias

404. Mr Neil A. **Sims**
Private Researcher
73-4369 Old Government, Mauka Road,
Kailua-Kona, HI 96740, USA
Contact No.(s) : (Tel./Fax) 3256516
Subject of Interest
Major : Bivalves, giant clams, marine aquaculture
Minor : Hatchery technology, statistics, stock assessment

405. Dr. Akhilesh K. **Singh**
Assistant Prof. Ranchi Veterinary College
Ranchi 834007, Bihar, India

Contact No.(s) : (Tel.) 455341
Subject of Interest
Major : Freshwater aquaculture, induced spawning integrated farming systems
Minor : Broodstock, hatchery technology, water quality
Subject of Consultancy
Freshwater aquaculture
Integrated fish farming
Carp hatchery construction.

406. Dr. C.S. **Singh**
Prof./Head of Fisheries, College of Fisheries
G.B. Pant University of Aquaculture and Technology
Pantnagar 263145, India
Subject of Interest
Major : Freshwater aquaculture, nutrition
Minor : Plankton

407. Dr. Vishwa Raman P. **Sinha**
Director, Central Institute of Fisheries Education
7 Bungalows, Versova, Bombay 400061,
India
Contact No.(s) : (Tel.) 6263404, 6261466
(Fax) 6261573
Subject of Interest
Major : Environmental impact, environmental studies, induced spawning
Minor : Carps, ecology, endocrinology

408. Mr. Dhurjati P. **Sinhababu**
Senior Scientist, Central Rice Research Institute
Cuttack 753006, Orissa, India
Contact No.(s) : (Tel.) 20020 L. 212
(Fax) 0671
Subject of Interest
Major : Freshwater aquaculture, reproductive physiology, rice-fish farming systems
Minor : Endocrinology, farming systems research, integrated farming systems

409. Dr. Mala P. **Sivalingam**
Prof., School of Biological Sciences
University Sains Malaysia
11800 Minden, Penang, Malaysia
Contact No.(s) : (Tel.) 877888 L. 3517
(Fax) 871526

Subject of Interest
Major : Biotechnology, environmental impact, environmental studies
Minor : Pollution, wastewater (sewage) reuse, water quality

410. Mr. Balasubramaniam **Sivaloganathan**
Research Scholar,
Deptt. of Zoology
National University of Singapore
Lower Kent Ridge, Singapore 0511
Contact No.(s) : (Tel.) 7722701
(Fax) 7792486

Subject of Interest
Major : Biochemistry, nutrition physiology
Minor : Bioenergetic, biotechnology, endocrinology

411. Mr. N. **Sivasankar**
Research Asstt.,
Regional Research Station
University of Agricultural Sciences
PB 24, Raichur, 584101, Karnataka State, India
Contact No.(s) : (Tel.) 8378

Subject of Interest
Major : Farming management, freshwater aquaculture, integrated farming systems
Minor : Brackishwater aquaculture, nutrition, pond management

412. Mr. Mike **Skladany**
Ph.D. Student, Sociology, Berkey Hall
Michigan State University, East Lansing
Michigan 48824-1111, USA
Contact No.(s) : (Tel.) 3555048

Subject of Interest
Major : Development, ecology, farming systems research, freshwater aquaculture, integrated farming systems, marine aquaculture, policy, resource economics, rice-fish farming systems.
Minor : Rural sociology

413. Mr. Allan H. **Smith**
Program Officer–Natural Resources
Caribbean Natural Resources Institute
Clark Street, Vieux Fort, St. Lucia
Contact No.(s) : (Tel.) 4546878
(Fax) 4545188
Subject of Interest
Major : Marine aquaculture, seaweeds
Minor : Coral Reefs

414. Dr. Prasert **Sobhon**
Prof. of Cell Biology, Anatomy Department
Faculty of Science, Mahidol University
Bangkok 10400, Thailand
Contact No.(s) : (Tel.) 2455198
(Fax) 2479880
Subject of Interest
Major : Biotechnology, cryopreservation, freshwater aquaculture
Minor : Biochemistry, parasites, reproductive physiology

415. Dr. William A. **Sodeman**, Jr.
Professor/Chief of Gastroenterology
Medical College of Ohia, PO Box 10008
Toledo, OH 43699-0008, USA
Contact No.(s) : (Tel.) 3813704
(Fax) 3820354
Subject of Interest
Major : Parasites
Minor : Public health

416. Mr. John D. **Sollows**
Box 1080, R.R. 2, Yarmouth
Nova Scotia, Canada B5A 4A6

Subject of Interest
Major : Extension, farming systems research, rice-fish farming systems
Minor : Cages, development, integrated farming systems

417. Dr. Patrick P. **Sorgeloos**
Director/Professor, Faculty of Agricultural Sciences
Laboratory of Aquaculture and Artemia Centre
Rozier 44, B-9000 Ghent, Belgium
Contact No.(s) : (Tel.) 643754
(Fax) 644193
Subject of Interest
Major : Hatchery technology, nutrition, seed production

418. Mr. Paul C. **Southgate**
Ph.D. Student, Zoology Department
James Cook University
Townsville, Queensland 4811, Australia
Contact No.(s) : (Tel.) 814530
(Fax) 251570
Subject of Interest
Major : Feed technology, marine aquaculture, nutrition
Minor : Biochemistry, bioenergetics, hatchery technology

419. Mr. Arunachala **Sreenivasan**
Fisheries Scientist, AE Block, Annanagar No. 45
Madras 600040, India
Contact No.(s) : (Tel.) 6212433
Subject of Interest
Major : Limnology, pollution, reservoirs
Minor : Environmental studies, freshwater aquaculture, water quality

420. Dr. Kostas I. **Stergiou**
Researcher, National Centre for Marine Research
Agios, Kosmas 16604, Helliniko,
Athens 16604, Greece
Contact No.(s) : (Tel.) 9821354
(Fax) 9522169

Subject of Interest
Major : Ecology, statistics, stock assessment
Minor : Farming management, integrated farming systems, marine aquaculture

421. Dr. Komar **Sumantadinata**
Head, Laboratory of Fish Breeding
Department of Aquaculture
Faculty of Fisheries, Institute Pertanian Bogor
Kampus IPB, Darmaga, Bogor, Indonesia
Contact No.(s) : (Tel.) 322167
Subject of Interest
Major : Genetics
Minor : Biotechnology, hatchery technology

422. Mr. Md. **Salequzzaman**
Associate Prof. and Head
Environment Science
Khulna University,
Khulna 9208, Bangladesh
Contact No.(s) : (Tel.) 880-41-720171-3 (ext. 265)
880-41-732154, 880-421-6053 (R)
(Fax) 880-41-732157, 880-41-732157 (R)
Subject of Interest
Major : Aquatic ecology Environmental impact assessment, Environmental monitoring and management.
Minor : Integrated Aquaculture, Fisheries management, Aquatic pollution, Air pollution, Landfall leachate and waste water management.
Subject of Consultancy
Environmental impact assessment of aquaculture projects. EIA on any types of projects, particularly solid waste management, air pollution, natural resources management etc. Integrated Aquaculture. Aquatic Ecology on mangrove sunderbans forest. Waste water aquaculture. Land fill leachates.

423. Mrs. Ismat Ara **Sayed**
Scientific Officer
Fisheries Research Institute
Kaowtkhali Mymensing-2207, Bangladesh
Contact No.(s) : (Tel.) 091-54874, 031-651561 (R)

Subject of Interest

Major : Aquaculture

Minor : Coastal Aquaculture and Environmental management.

Subject of Consultancy

Aquaculture management. Natural shrimp farming system. Reuse of organic waste, duckweed production and its application.

424. Dr. Ranjit **Singh**
Asst. Zoologist (Fish)
Deptt. of Fisheries, P.A.U.,
Ludhiana 141004, India
Contact No.(s) : (Tel.) 0161-401960 (Ext. 432)
0161-476985 (R)
(Fax) 0161-400945

Subject of Interest

Major : Fish nutrition and feed formulation. Intensive and semi intensive aquaculture.

Minor : Fish breeding and ecology.

Subject of Consultacny

Semi-intensive and intensive fish culture. Feed formulation. Fish farm establishment. Hatchery establishment, Breeding techniques, Running water fish culture.

425. Dr. H. **Shivanandamurthy**
Associate Professor,
Aquaculture College of Fisheries
University of Agricultural Sciences
Mangalore-575002, India
Contact No.(s) : (Tel.) 082-439256, 434356
08255-23166 (R)
(Fax) 0824-438366

Subject of Interest

Major : Fresh water prawn hatchery production, fresh water prawn and carp farming, Aquaculture nutrition and feed technology.

Minor : Farming system of prawn and fish, shrimp farming and seed production, Project formulation and implementation.

Subject of Consultancy
Fresh water prawn hatchery construction and seed production. Feed development for prawn and fish larvae. Feed development for freshwater prawn, carps and shrimps in nursery and grow out farming. Prawn, shrimp, and carp farm design and project preparation. Aquaculture nutrition and feed technology. Management of freshwater prawn, fish and shrimp culture. Use of sugar cane factory / distillary waste for fish and prawn culture

426. Mr. Arul V. **Suresh**
Ph.D. Student
Cooperative Fisheries Research Laboratory
Southern Illinois University Carbondale,
IL 2901, USA
Contact No.(s) : (Tel.) 5367761
(Fax) 5367761
Subject of Interest
Major : Freshwater aquaculture
Minor : Development, production systems

427. Dr. James P. **Szyper**
Researcher, Hawaii Institute of Marine Biology
University of Hawaii at Manoa
PO Box 1346, Kaneohe, HI 96744, USA
Contact No.(s) : (Tel.) 2367420
(Fax) 2367443
Subject of Interest
Major : Pond dynamics, waste-fed, water quality
Minor : Ecology, engineering phytoplankton

428. Dr. Albert G.J. **Tacon**
Fishery Resources Officer, Rm. F-227, FIRI,
Fisheries Department, FAO
Viale delle Terme di Caracalla
00100, Rome, Italy
Contact No.(s) : (Tel.) 57976470
(Fax) 5120330

Subject of Interest
Major : Feed technology, nutrition, waste-fed
Minor : Farming systems research, integrated farming systems, pond management

429. Mr. Andre J. **Talbot**
Research Associate
University de Quebec A Chicoutimi
555 Boul. de University, Chicoutimi
Quebec, Canada G7H 2B6
Contact No.(s) : (Tel.) 6503618
(Fax) 6503419
Subject of Interest
Major : Statistics
Minor : Quantitative genetics

430. Mr. Rashid A. **Tamatamah**
Fisheries Officer
Marine Fisheries Research and Training Institute
PO Box 60091, Dar Es Salaam, Tanzania
Contact No.(s) : (Tel.) 47011 to 3
Subject of Interest
Major : Marine aquaculture, pond dynamics, site selection
Minor : Farming systems research, nutrition, stock assessment

431. Dr. Eddy S.P. **Tan**
Assoc. Prof. School of Biological Sciences
University Sains Malaysia
11800 Minden, Penang, Malaysia
Contact No.(s) : (Tel.) 877888 L. 3534
(Fax) 871526
Subject of Interest
Major : Behavior, endocrinology, reproductive physiology
Minor : Carps, freshwater aquaculture, seed production

432. Mr. Touch Seang **Tana**
Administrator, Deptt. of Fisheries
N-186 134 Tou Samuth, P. Pesah, Cambodia
Contact No.(s) : (Tel.) 23275

Subject of Interest
Major : Environmental impact, information, marketing
Minor : Brackishwater aquaculture, freshwater aquaculture, rural sociology

433. Mr. Hideyuki **Tanaka**
Project Manager
FAO South Pacific Aquaculture Development Project
C/o UNDP Private Mail Bag, Suva, Fiji
Contact No.(s) : (Tel.) 300067
(Fax) 300029
Subject of Interest
Major : Brackishwater aquaculture, freshwater aquaculture marine aquaculture
Minor : Development, farming management, marketing

434. Dr. Nobuhiko **Taniguchi**
Prof. Department of Cultural Fisheries
Faculty of Agril., Kochi University
Monobe B-200, Nankoku, Kochi 783, Japan
Contact No.(s) : (Tel.) 634141
(Fax) 636492
Subject of Interest
Major : Genetic resources, population genetics, quantitative genetics
Minor : Broodstock, carps, finfish, seed production

435. Ms. Mohosena B. **Tanu**
Scientific Officer, Fisheries Research Institute
Chandpur 3602, Bangladesh
Contact No.(s) : (Tel.) 3407, 3482
Subject of Interest
Major : Nutrition
Minor : Crustaceans, freshwater aquaculture

436. Mr. Moh **Taufiq**
JI, Jombang IV/1, BTN Taman Baru
Mataram, Indonesia
Contact No.(s) : (Tel.) 26906

Subject of Interest

Major : Farming management, production systems, resource economics, seaweeds, socioeconomics, statistics

Minor : Brackishwater aquaculture environmental studies, rice-fish farming systems

437. Mr. Shawn R. **Taylor**

Senior Biologist, 158 Mountain View Road,
S. Georgetown, Ontario, Canada L7G 4K7
Contact No.(s) : (Tel.) 2294646
(Fax) 2294692

Subject of Interest

Major : Crustaceans, marine aquaculture, wastewater (sewage) reuse

Minor : Integrated farming systems, pond dynamics, rural sociology

438. Ms. Saidin **Thalathian**

Fisheries Research Officer
Freshwater Fish Research Centre
Deptt. of Fisheries, Batu, Berendam
75300 Melaka, Malaysia
Contact No.(s) : (Tel.) 353497, 353499
(Fax) 355705

Subject of Interest

Major : Hatchery technology, induced spawning, site selection

Minor : Freshwater aquaculture, post-harvest technology, production systems

439. Dr. Hugh W. **Thomforde**

Pearl Farm Manager
Tongareva Marine Research Centre
Ministry of Marine Resources, PO Box 85, Avarua
Rarotonga, Cook Islands
Contact No.(s) : (Tel.) 28721 to 3
(Fax) 29721

Subject of Interest

Major : Bivalves, water quality

Minor : Farming management

440. Dr. John E. **Thorpe**
Senior Principal Scientific Officer
Agril. and Fisheries Department
The Scottish Office
Freshwater Fisheries Laboratory, Faskally
Pitlochry, Perthshire PH16 5LB, UK
Contact No.(s) : (Tel.) 472060
(Fax) 473523
Subject of Interest
Major : Development, ecology, genetic resources
Minor : Behavior, growth, reproductive physiology

441. Dr. Dirk **Thys van den Audenaerde**
Director, Royal Museum for Central Africa
Steenweg op Leuven, 13-3080, Tervuren,
Belgium
Contact No.(s) : (Tel.) 7695285
(Fax) 7670242
Subject of Interest
Major : Behavior, taxonomy, tilapias
Minor : Catfishes, freshwater aquaculture

442. Dr. Ruangrai **Tokrisna**
Associate Professor
Deptt. of Agril. and Resource Economics
Kasetsart University, Bangkok 10900, Thailand
Contact No.(s) : (Tel.) 5613467
(Fax) 5798781
Subject of Interest
Major : Economics
Minor : Marketing, resources economics

443. Mr. Kamonporn **Tonguthai**
Director, Aquatic Animal Health
Research Institute
Kasetsart University Campus
Bangkok 10900, Thailand
Contact No.(s) : (Tel.) 5794122
(Fax) 5613993

Subject of Interest
Major : Diseases, parasities, pathology
Minor : Water quality

444. Dr. Jorge E. **Toro**
Instituto de Biologia
Marina Casilla 567, Valdivia, Chile
Contact No.(s) : (Fax) 212953
Subject of Interest
Major : Genetic resources, marine aquaculture, quantitative genetics
Minor : Broodstock, population genetics, production systems

445. Ms. Krisna R. **Torrissen**
Senior Scientist, Directorate of Fisheries
N-5198 Matredal, Norway
Contact No.(s) : (Tel.) 366040
(Fax) 366143
Subject of Interest
Major : Biochemistry, growth, nutrition
Minor : Biotechnology, genetics, population genetics

446. Mr. Alamoussa C. **Traora**
Directeur des peches
Ministere de I'Environnement et du Tourisme
03 BP 7044, Ouagadougou 03, Burkina Faso
Contact No.(s) : (Tel.) 306394, 306398
Subject of Interest
Major : Integrated farming systems, limnology, tilapias
Minor : Environmental impact, farming systems research, water quality

447. Mr. Truong Quoc **Phu**
Prof., Faculty of Fisheries
Cantho University, Cantho, Vietnam
Subject of Interest
Major : Biotechnology, marine aquaculture
Minor : Computer methods

448. Dr. Paul J. **Udo**
Research Fellow I
Deptt. of Fisheries and Aquaculture
University of Calabar, PMB 1115, Calabar
Cross River State, Nigeria
Subject of Interest
Major : Crustaceans, integrated farming systems
Minor : Freshwater aquaculture

449. Dr. Adiaha A.A. **Ugwumba**
Lecturer, Department of Forestry
Fisheries and Wildlife Management
University of Agriculture
Abgokuta, Nigeria
Contact No.(s) : (Tel.) 200170 to 177
Subject of Interest
Major : Freshwater aquaculture, induced spawning, nutrition
Minor : Environmental studies, extension, post-harvest technology

450. Dr. Alex O. **Ugwumba**
Lecturer, Department of Zoology
University of Ibadan, Ibadan, Nigeria
Contact No.(s) : (Tel.) 400550 to 62
Subject of Interest
Major : Freshwater aquaculture, induced spawning, nutrition
Minor : Conservation, environmental studies, pollution

451. Mr. Graham F. **Usher**
Nature Conservation Officer
National Resources Management Project
PO Box 1375, Manado, Sulut, Indonesia
Contact No.(s) : (Tel.) 64296
(Fax) 64296
Subject of Interest
Major : Conservation, coral reefs, giant clams
Minor : Marine aquaculture, seaweeds

452. Dr. Fred M. **Utter**
Acting Director, School of Fisheries
University of Washington, Seattle, WA 98195, USA
Subject of Interest
Major : Conservation, genetic resources, population genetics
Minor : Hatchery technology, pens, taxonomy

453. Mr. Jan Michael **Vakily**
Research Scientist, ICLARM
C/o EC Delegation Sierra Leone, Valise Diplomatique
Contact No.(s) : (Tel.) 250670
(Fax) 225212
Subject of Interest
Major : Stock assessment
Minor : Brackishwater aquaculture, computer methods

454. Mr. Anne **van Dam**
Thorbechestraat 60, 6702, Bs Wageningen
The Netherlands
Subject of Interest
Major : Bioenergetics, modeling, pond dynamics
Minor : Ecophysiology, food-chains, growth

455. Mr. Frans **van den Gerg**
Aquaculture, C/o FAO, PO Box 30750
Lilongwe 3, Malawi
Subject of Interest
Major : Extension, freshwater aquaculture, integrated farming systems

456. Mr. Wim L. **van Densen**
University of Fish Culture and Fisheries
Wageningen Agril University, PO Box 338
6700 AH Wageningen, The Netherlands
Contact No.(s) : (Tel.) 83307
(Fax) 83962
Subject of Interest
Major : Foodchains, reservoirs, tilapias

457. Mr. Peter G.M. **van der Heijden**
Aquaculturist, C/o De Kempen 3
5345 NJ Oss, The Netherlands
Subject of Interest
Major : Extension, freshwater aquaculture, tilapias
Minor : Integrated farming systems, production systems, rice-fish farming systems, stock assessment, waste-fed

458. Mr. Joxef J.G. **van der Zanden**
Project Coordinator
Institute of Animal Breeding and Genetics
Albrecht-Thaer-Weg 1, 3400 Gottingen,
Germany
Contact No.(s) : (Tel.) 395604
Subject of Interest
Major : Brackishwater aquaculture, farming systems research, marine aquaculture
Minor : Cages, catfishes, crustaceans

459. Dr. Charles **van Hove**
Prof. Plant Biology Laboratory
Catholic University of Louvain
5, Place Croix du Sua
Bte 14 1348 Louvain-la-Neuve, Belgium
Contact No.(s) : (Tel.) 473464 to 8
(Fax) 473471
Subject of Interest
Major : Integrated farming systems, rice-fish farming systems

460. Dr. Jan H. **van Weerd**
Researcher, Deptt. of Fish culture and Fisheries
Wageningen Agril. University, PO Box 338
6700 AH Wageningen, The Netherlands
Contact No.(s) : (Tel.) 83307
(Fax) 83962
Subject of Interest
Major : Freshwater aquaculture, integrated farming systems, reproductive physiology
Minor : Brackishwater aquaculture, environmental studies

461. Mr. Koen A. **Vanlerberghe**
Lecturer, Zoology Department
Kenyatta University, Box 43844,
Nairobi, Kenya
Contact No.(s) : (Tel.) 810901 L. 305, 307
Subject of Interest
Major : Farming management, freshwater aquaculture, integrated farming systems.
Minor : Carps, catfishes, tilapias

462. Dr. Tharayil J. **Varghese**
Prof., College of Fisheries
Mangalore 575002, India
Contact No.(s) : (Tel.) 27834
Subject of Interest
Major : Carps, feed technology, freshwater aquaculture
Minor : Brackishwater aquaculture, hatchery technology marine aquaculture

463. Mr. Syama S. **Veerina**
C/o Dr. V. Kesava Rao
Wahab Road, Tenali, A.P. 522201, India
Subject of Interest
Major : Brackishwater aquaculture, freshwater aquaculture, nutrition
Minor : Computer methods, modeling, statistics

464. Mr. Juan Andres **Vega Durand**
Asesor, Dpto. Ecologia y Recursos Naturales
Universidad Centroamerica
Apartado 69 Managua, Nicaragua
Contact No.(s) : (Tel.) 780951
(Fax) 624004 to 5
Subject of Interest
Major : Farming systems research, integrated farming systems, tilapias
Minor : Computer methods, engineering, statistics

465. Ms. Ravelina R. **Velasco**
Research Assistant, ICLARM
MCPO Box 2631, Philippines

Contact No.(s) : (Tel.) 8189283, 8180466, 8175255
(Fax) 8163183
Subject of Interest
Major : Freshwater aquaculture, quantitative genetics
Minor : Cryopreservation, reproductive physiology, seed production

466. Mr. T.S. **Velayudhan**
Scientist, Central Marine Fisheries Research Institute
PB No. 2704, Cochin 682031, India
Contact No.(s) : (Tel.) 369867 L. 238
Subject of Interest
Major : Bivalves, coral reefs growth
Minor : Biotechnology, induced spawning, site selection stock assessment

467. Dr. John A.J. **Verreth**
University Lecturer
Deptt. of Fish Culture and Fisheries
Wageningen Agril. University, PO Box 338
6700 AH Wageningen, The Netherlands
Contact No.(s) : (Tel.) 83307
(Fax) 83962
Subject of Interest
Major : Bioenergetics, catfishes, nutrition
Minor : Farming systems research, freshwater aquaculture, pond dynamics

468. Dhananjay Kr. **Verma**
Central Institute of Freshwater Aquaculture
Kausalyaganga, Bhubneshwar, Orissa
Subject of Interest
Major : Fisheries
Minor : Aquaculture
Subject of Consultancy
Fish breeding, Pond farm management, Integrated fish farming. Freshwater prawn culture management.

469. Mr. Niraj Kumar **Verma**
Qr. 2K/95, Bariatu Housing Colony
Ranchi-834009, India
Contact No.(s) : (Tel.) 0651-542831 (R)
Subject of Interest
Major : Freshwater prawn and marine prawn culture and their hatchery
Minor : Freshwater fish culture prospects
Subject of Consultancy
Freshwater prawn culture and hatchery operation.
Marine prawn culture and hatchery operation.
Breeding of ornamental fishes.

470. Dr. Janzel R. **Villalaz**
Research Scientist
Centro de Ciencias del Mar Y Limnologia
Universidad de Panama
Cuidad Universitaria "Octavio Mendez Pereira"
Estafeta Universitaria, Panama, Panama
Contact No.(s) : (Tel.) 284409
(Fax) 237870
(E-mail : VILLALAZ.HURACAN)
Subject of Interest
Major : Bivalves, computer methods, genetics

471. Dr. Wolfgang **Villwock**
University Professor
Zoologisches Institut und Zoologisches Museum
Universitat Hamburg, Martin Luther King PL. 3
D-2000 Humburg 13, Germany,
Contact No.(s) : (Tel.) 41233873
(Fax) 41233937
Subject of Interest
Major : Environmental impact, genetic resources
Minor : Freshwater aquaculture, tilapias

472. Dr. Wade O. **Watanabe**
Chief Scientist,
Caribbean Marine Research Centre
805 East 46th PL, Vero Beach, FL 32963,
USA

Contact No.(s) : (Tel.) 2349931
(Fax) 2349954

Subject of Interest

Major : Brackishwater aquaculture, finfish, marine aquaculture

Minor : Hatchery technology, induced spawning, tilapias

473. Mr. Ian G. **Watson**
Feed Tachnologist, Natural Resources Institute
Central Avenue, Chatham Maritime, Chatham
Kent ME4 4TB, UK
Contact No.(s) : (Tel.) 883552
(Fax) 880066, 880077

Subject of Interest

Major : Development, extension, feed technology

Minor : Environmental impact, pesticides

474. Dr. Leong K. **Wee**
Senior Lecturer, Key Centre for Aquaculture
University of Tasmania, PO Box 1214
Launceston, Tasmania 7250, Australia
Contact No.(s) : (Tel.) 260538
(Fax) 260236

Subject of Interest

Major : Feed technology, finfish nutrition

Minor : Freshwater aquaculture, marine aquaculture, tilapias

475. Dr. Richard S.J. **Weisburd**
Fellow, National Institute for Environmental Studies
Global Environment Division, 16-2 Onogawa
Tsukuba, Ibaraki 305, Japan
Contact No.(s) : (Tel.) 516111 L. 327
(Fax) 564680

Subject of Interest

Major : Ecology, pond dynamics, water quality

Minor : Computer methods, detritus, integrated farming systems

476. Dr. Robin **Welcomme**
Chief, Inland Water Resources and Aquaculture
Service, FAO-FIRI,
Rm. F-237, Viale delle Terme di Caracalla
00100 Rome, Italy
Contact No.(s) : (Tel.) 57976654
(Fax) 5120330
Subject of Interest
Major : River fisheries, species introduction
Minor : Fishery management

477. Dr. M. Jayantha S. **Wijeyaratne**
Assoc. Prof. Department of Zoology
University of Kelaniya, Kelaniya, Sri Lanka
Contact No.(s) : (Tel.) 521916
Subject of Interest
Major : Environmental impact, environmental studies, lagoons,
Minor : Ecology, pollution, reservoirs

478. Mr. Scott A. **Willis**
Research Scientist, Florida Marine Research Institute
100 8th Ave., S.F., St. Petersburg,
FL 33701, USA
Contact No.(s) : (Tel.) 8968626
(Fax) 8230166
Subject of Interest
Major : Environmental studies, marine aquaculture, stock assessment
Minor : Ecology, genetics, population genetics

479. Dr. Giora W. **Wohlfarth**
Research Scientist
Fish and Aquaculture Research Station
Dor, D.N. Hof Hacarmel 30820, Israel
Contact No.(s) : (Tel.) 390651
Subject of Interest
Major : Carps, genetics, tilapias

480. Dr. **Wong Tatt Meng**
Director, Centre for Marine and Coastal Studies
Universiti Sains Malaysia
11800 Minden, Penang, Malaysia
Contact No.(s) : (Tel.) 877888 L. 2103
(Fax) 872960
Subject of Interest
Major : Bivalves, induced spawning, seed production
Minor : Marine aquaculture, post-harvest technology, production systems

481. Dr. Jahara **Yahaya**
Associate Professor
Faculty of Economics & Administration
University of Malaya, 59100 KL, Malaysia
Contact No.(s) : (Tel.) 7554111
(Fax) 7567252
Subject of Interest
Major : Resource economics
Minor : Economics

482. Dr. Ahmed **Yahyaoui**
Enseignant-chercheur
Deptt. de biologie, Faculte des Sciences
BP 1014, Rabat, Morocco
Contact No.(s) : (Tel./Fax) 7775461
Subject of Interest
Major : Brackishwater aquaculture, computer methods
Minor : Ecology, genetic resources, growth

483. Dr. Masashi **Yamaguchi**
Prof., Department of Marine Sciences
University of the Ryukyus, Senbaru 1
Nishihara, Okinawa 903-01, Japan
Contact No.(s) : (Tel.) 8952221 L. 2685
(Fax) 8952247
Subject of Interest
Major : Bivalves, fisheries enhancement
Minor : Conservation, coral reefs, lagoons

484. Mr. Takashi **Yamamoto**
Okinawa Prefectural Fisheries Experimental Station
Yaeyama Branch 828-2 Kabira
Ishigaki-shi, Okinawa 907-04, Japan
Subject of Interest
Major : Marine aquaculture, seed production, stock assessment
Minor : Behavior, finfish

485. Dr. Alejandro **Yanez-Arancibia**
Scientific Director
Av, Agustuin Melgar y Juan de la
Barrera Cd. Universitaria
Apartado Postal 520, Codigo Postal 24030,
Mexico
Contact No.(s) : (Tel.) 66589
(Fax) 65954
Subject of Interest
Major : Ecology, environmental impact, lagoons

486. Dr. Chia V. **Yapi**
Attache de Recherche, Department des ressources
Institut des Savanes, 01 BP 633, Bouake 01
Cote d'Ivoire
Contact No.(s) : (Tel.) 633364
(Fax) 632045
Subject of Interest
Major : Freshwater aquaculture, quantitative genetics
Minor : Computer methods, genetic resources, statistics

487. Mr. Ibrahim **Yaro**
Deputy Chief Fisheries Officer,
Fisheries Department
Ministry of Agrilcultural and Natural Resources Minna
PMB 74, Minna, Niger State, Nigeria
Subject of Interest
Major : Freshwater aquaculture, reservoirs, seed production
Minor : Conservation, hatchery technology, induced spawning, integrated farming systems, policy, rice fish farming systems

488. Dr. Shinn-Pyng **Yeh**
Assoc. Prof., Department of Aquaculture
National Ping-Tung Institute of Agriculture
1 Hseuh-Fu Road, Nei-Pu Husiang, Pingtung
Taiwan 92107, China
Contact No.(s) : (Tel.) 7703202 L. 2902
(Fax) 7700173
Subject of Interest
Major : Ecology
Minor : Environmental studies, finfish, taxonomy

489. Dr. Tientcheu J. **Youmbi**
Fisheries and Oceanography Research Station
PMB 77, Limbe, Cameroon
Subject of Interest
Major : Brackishwater aquaculture, farming management marine aquaculture
Minor : Feed technology, nutrition

490. Mr. Yuan **Zhanpeng**
Research Assoc. Institute of Reservoir Fisheries
Ministry of Water Resources, Academia Sinica
Wuhan 430073, China
Contact No.(s) : (Tel.) 703023
Subject of Interest
Major : Genetics
Minor : Biotechnology, reservoirs

491. Mr. Sutandar **Zainal**
Lecturer Staff, Department of Fisheries
Padjadjaran University, Jatinangor 4563,
Indonesia
Contact No.(s) : (Tel.) 443998
Subject of Interest
Major : Freshwater aquaculture
Minor : Cages, reservoirs

492. Mr. Ronald D. **Zweig**
Aquaculturist, The World Bank, 1818 H Street
NW, Washington DC 20433, USA

Contact No.(s) : (Tel.) 4582174
(Fax) 4771865

Subject of Interest

Major : Economics, freshwater aquaculture, integrated farming systems

Minor : Limnology, pond dynamics, reservoirs

493. Mrs. Roseemma Wamaa **Enisuamensah**
Water Research Institute CSIR
PO Box 38, Achimota
Ghana, West Africa
Contact No.(s) : (Tel.) 775511, 761031
(Fax) 233-21-761030

Subject of Interest

Major : Coastal fishery management.

Minor : Mangroves, lagoon fisheries

Index

F

G

QUESTIONNAIRE

1. Name ☐ Prof. ☐ Dr. ☐ Mr. First Middle Last

2. Year of Birth 3. Nationality 4. Sex

5. Official Address

6. Private Address

Phone (s)

Fax

E-mail

Phone (s)

Fax

Tlx

7. Subject (s) Topic (s) of interest.

Major :

Minor :

8. Please specify, if your currently engaged in any research on Fisheries / Aquaculture.

9. Area in which you would like to provide consultancy.

Date | Signature

(Xerox Copy of form can also be used)

Please return to : **Daya Publishing House,**
1123/74, Deva Ram Park,
Tri Nagar, Delhi-110035 (INDIA)
Phone : 710 3999
Or Fax : (011) 719 8902
E-mail : info@dayabooks.com
Website : www.dayabooks.com